STRATEGIC GIS PLANNING AND MANAGEMENT IN LOCAL GOVERNMENT

T0199564

STRATEGIC GIS PLANNING AND MANAGEMENT IN LOCAL GOVERNMENT

David A. Holdstock

CRC Press
Taylor & Francis Group
Boca Raton London New York

CRC Press is an imprint of the
Taylor & Francis Group, an **informa** business

CRC Press
Taylor & Francis Group
6000 Broken Sound Parkway NW, Suite 300
Boca Raton, FL 33487-2742

First issued in paperback 2019

ISBN-13: 978-1-4665-5650-8 (hbk)
ISBN-13: 978-0-367-86740-9 (pbk)

Library of Congress Cataloging-in-Publication Data

Names: Holdstock, David A., author.
Title: Strategic GIS planning and management in local government / David A. Holdstock.
Description: Boca Raton, FL : CRC Press, 2017. | Includes bibliographical references and index.
Identifiers: LCCN 2016020195 | ISBN 9781466556508 (alk. paper)
Subjects: LCSH: Geographic information systems. | Local government--Technological innovations.
Classification: LCC G70.212 .H65 2017 | DDC 352.3/80285--dc23
LC record available at https://lccn.loc.gov/2016020195

Visit the Taylor & Francis Web site at
http://www.taylorandfrancis.com

and the CRC Press Web site at
http://www.crcpress.com

To my extraordinary wife; to my daughters Natalie, Amelia, and Bea; and to my son Tosh. This book is for you because you have all been so unbelievably supportive.

Contents

Preface

"A lot of people in our industry haven't had very diverse experiences. So they don't have enough dots to connect, and they end up with very linear solutions without a broad perspective on the problem. The broader one's understanding of the human experience, the better design we will have."

Steve Jobs

I consider mine a life that was well lived.

Whether it's sailing in the Bahamas, salmon fishing in Alaska, taking hot air balloon rides over Dubai, or taking nerve-wracking business trips to West Africa, I've had some extraordinary experiences in this world. Some were good; some were bad, but all of them were made possible because of my career in geographic information systems (GIS).

Supported by a great business partner and fueled by a burning desire to succeed, I've spent nearly half of my life planning, designing, and deploying GIS. Since we started Geographic Technologies Group (GTG) in 1997, we've enjoyed overwhelming success in securing GIS strategic implementation planning work all over the world, predominantly in the United States.

Consider this a playbook. It is a formula on how to plan, manage, and maintain GIS as told through the lens of one man's two-and-a-half decade journey through this maturing industry. Unlike other authors who have written on this subject, I am a businessman, not an academic nor an intellectual. I have spent the last 25 years talking to and listening to local government professionals, watching them as they build consensus. I have organized and orchestrated too many Blue Sky sessions to count and have performed hundreds of GIS requirement analyses. I have documented government business processes and workflow and a detailed enterprise system design and system architecture and, most importantly, I have built a working relationship with local government.

Our record in the industry seems to be unmatched. To deploy a baseball metaphor, I think that we are batting at 1000. We've worked successfully in government agencies small and large, everywhere from towns and villages to fishing ports and metropolitan areas. To win this kind of work, we write proposals and develop graphic-rich presentations that must not only explain the proposed implementation and management strategies but also resonate with the towns, cities, and counties that form the GIS community.

This may be a story of my journey, but without the influence and support of a very organized and talented business partner, Mr. Curt Hinton, along with that of a long-time client, friend, innovator in government, and key editor of this book, Ms. Cathy Raney, this book would not have been possible.

As for me, I come from a Yorkshire fishing town in the north of England. My hometown was so insular and provincial that when, on April 15, 1912, after the RMS *Titanic* capsized on its maiden voyage to New York, the headlines in our local newspaper, the *Yorkshire Gazette*, read, "Two Yorkshire Men Lost at Sea."

It goes without saying that all 26 of the *formative* years I've spent in Yorkshire and Nottinghamshire were fairly provincial, especially when it came to understanding people and technology. Now, I have 23 years as a North Carolinian and 3 years as a New Yorker beneath my belt. I'm pleased to report that my cultural habits have changed rather drastically: I'm as comfortable eating half a dozen Krispy Kreme donuts as I am taking a covered dish to a *pig pickin'*. At 53 years of age, I have lived almost exactly half my life in the United Kingdom and half in the United States.

British people are perpetually ready for humor. Our food may be bland, but irony seasons even our most serious conversations. The British mode of joking but not joking, caring but not caring, and being serious without being serious can get frustrating at times, so I apologize to the readers if my musings are in any way ambiguous or grating. Think of it this way: I am being sincere, without, perhaps, being serious, although admittedly, Taylor & Francis may have had more success asking the Marx Brothers to write this book. But they're hard to reach these days, so here I am.

I have to warn you that I relied upon two very important techniques while writing this book: first, while writing, I poured myself a healthy glass of truth serum, which I buy in bottles at the grocery store named Cabernet, Merlot, and Malbec.

Secondly, I play the *obvious game*, where I offer fairly self-evident advice about the various stages of GIS development, just to make sure that I have all of the bases covered. This book is born from my experiences proposing, presenting, and developing over 200 GIS strategic implementation plans. I'm confident that this book will interest and amuse you and convince you that a formula for success may just be possible.

My biggest vote of thanks goes to my parents for bringing me into a very British world. In all seriousness, what I learned from my parents was that to live a purposeful, blessed, and happy life, we need to stay on our game. We need to stay relevant and work hard because what we do in this life may just well echo through eternity!

David A. Holdstock

Acknowledgments

If you ask me how I could possibly acknowledge every single person who has contributed to this book, I would have to say, unfortunately, that is where I run out of words. It may be just one of those impossible tasks because there have been so many people who have influenced and helped me along the way.

My high school geography teacher, Ms. Ursula Hammond at Bingham Toothill Comprehensive in Nottinghamshire, had something to do with my initial interest in human settlement patterns. So, Mrs. Hammond, thank you. My undergraduate professors at Thames Polytechnic certainly established a foundation of enthusiasm, and an exceptional and incredibly considerate GIS professor at North Carolina State University, Dr. Hugh Devine, may have helped choreograph my career. Mr. Curt Hinton, my gifted and talented business partner of 20 years, whose tolerance and understanding has allowed us to grow and enjoy GTG, must receive a serious portion of gratitude.

Some authors have certainly influenced me to write this book. Thanks to Bill Bryson for reminding me that I didn't grow up where there were glow-in-the-dark insects, and that British people of all ages and social backgrounds can still get excited about a warm beverage. That would be called a cup of tea. Other great writers with style and purpose need to be recognized like Daniel Gilbert, Ken Robinson, Geoff Colvin, Adam Alter, Heather Campbell and Ian Masser, Bill Gates, and Les Dawson.

My heartfelt thanks goes to all the local government professionals whom I have had the fortune to work with over the last 25 years. Please see Appendix A for a list of the local government organizations.

And to use my children's vernacular, a *shout-out* goes to a person who made me stronger: my friend Ms. Cathy Raney, GIS coordinator, Campbell County, Wyoming, who convinced me that this book was a noble cause. There is a young gifted man named Cameron Higgins, whose original thinking and ability to marshal my facts was an integral part of producing this book. Toward the end of the book, he spoke to me exclusively in warnings. Thank you Cam! Also, a big thanks goes to Ms. Brittany Taylor, whose exceptional design talent and unbelievable keen eye produced the graphic rich artwork for the book.

And finally, a big thanks goes to the smartest, most considerate, and most talented civil engineer whom I have had the fortune of meeting—my wife. She is the visible personification of perfection. I will love her until God shall separate us by death.

David A. Holdstock

Acknowledgment

Author

Mr. David Holdstock, BA, MS, a geographic information systems (GIS) professional and chief executive officer, co-established and incorporated Geographic Technologies Group, Inc. in 1997. Mr. Holdstock is a GIS practitioner and a leading expert in developing enterprise and sustainable GIS strategic implementation plans for towns, cities, and counties. Over the past 25 years, Mr. Holdstock has planned, designed, and coordinated the adoption and implementation of GIS technology for over 200 local government organizations. He has published many articles on GIS strategic planning for local government, including assessing the value of GIS; understanding the challenges, barriers, and pitfalls of GIS; and the implementation process for local government. David has conducted hundreds of workshops, seminars, and discussions on GIS implementation. His previous work experience has included being a GIS manager for the world's leading transportation engineering company in New York and a GIS director for a research institute at North Carolina State University.

List of Figures

1

Introduction

A goal without a plan is just a wish.

Antoine de Saint-Exupery

1.1 Background

We are about to enter a new world of advanced and innovative spatial analytics in local government and, with it, a future that includes a very different relationship with geospatial technology. We must change the way we train, educate, and transfer GIS knowledge to one another. We must understand what drives people to learn, think, and innovate. We must find new ways of capturing and appreciating outstanding human performance. All of this will include completely different management and learning styles and a new generation of Geographic Information Officers (GIO). Your goal is to plan, design, and build an enduring, sustainable, and enterprise GIS that first asks the questions, how healthy is my organization now, and how can we be prepared for the future? The good news is that a corporate-style diagnosis in the form of this book's formula for success is the starting point for all local governments. Knowing your strengths and weaknesses will allow you to prioritize investment and resources.

Are you prepared to take on this multiyear challenge?

The adoption of geospatial technologies within local government organizations has increased the demand for sophisticated planning tools and techniques to assist in the complex implementation process. As geospatial technology becomes ever more societally relevant, government organizations are beginning to understand the real opportunities as well as the complexities that this technology presents. GIS exists not only in the office but also on every smartphone and mobile device.

Currently, the real question is, what underpins a successful, enduring, and enterprise GIS in local government? Moreover, what *is* an enterprise, sustainable, and enduring GIS? In short, a sustainable and enduring *enterprise GIS* describes an integrated solution that serves an entire organization by offering (a) levels of geospatial functionality, (b) uniform standards, (c) good governance, (d) reliable digital data and databases, (e) workflow procedures, (f) training education and knowledge transfer, and (g) a backbone for architecture and infrastructure.

I cannot say how the future will unfold. What I can say is that throughout the past two decades, I have seen local governments not only repeat many of the same mistakes but also make the same mistakes at the same time.

I know that implementing new, innovative software solutions is not always easy. If local governments lack perspective on the effective use of GIS, the task of implementation can be cumbersome and time consuming. Deploying a structured and comprehensive strategic plan can introduce positive changes into the way that the GIS is adopted and embraced.

Everyone wants to make a difference in their community. For the GIS and government professionals to do so, solutions to the technical, tactical, logistical, and political problems that they face must be found and acted upon. Municipalities need solutions that will put their organization on the map.

I wrote this book to make a set of tools that are available to the local government professionals and GIS communities. I am hoping that if these tools are utilized correctly, they will allow your organization to develop a sustainable and effective relationship with geospatial technology. I call this toolkit the formula for success and hope that it can help you as much as it has helped me. But you may be wondering, "David, what *exactly* will this book give me?" Well, formally, it will provide you the following:

- A GIS formula for evaluating, benchmarking, and implementing an enterprise GIS
- An explanation of the challenges, barriers, and pitfalls of GIS implementation
- A strategy for developing a GIS vision and goals and objectives for an enterprise GIS
- An understanding of the importance of GIS governance and governance models in local government
- An enterprise GIS training, education, and knowledge transfer solution
- A thoughtful approach to quantifying and qualifying the benefits of GIS technology
- How to sell GIS

Understanding the principles that underpin success in the GIS community requires, more than anything, common sense. Informally, my book aims to help you with everything that falls outside of that particular realm. I am not just interested, however, in making the implementation and utilization of

GIS technology that much easier but also in showing geospatial technology for what it is: a dynamic and ever-evolving tool for problem-solving. That is why GIS is so socially relevant and why, by using a combination of theory and real-world practice, this book provides invaluable insight into the implementation and governance of sustainable GIS technologies.

To achieve these aims, I emphasize the importance of both strategic planning and having a firm understanding of the potential for GIS as a technological innovation. I demonstrate how one develops goals and objectives, and presents the benefits and costs of GIS, along with the methods to achieve and present measurable results.

Over the next few chapters, you will learn everything, from how to secure buy-in from elected officials and senior management; where to identify funding and investment strategies; and how to eliminate data inaccuracies, redundancies, and duplication.

Using theory and real-world practice, this book offers a perspective on the process of implementation. It discusses recent innovations in GIS technology and contrasts the contemporary corporate, cloud-based style of GIS implementation with older styles of adoption and their constraints. Most importantly, this book emphasizes the value of strategic planning and an understanding of GIS as a multifaceted technological innovation.

This book will show how to overcome the barriers to successful implementation by detailing all of the procedural components that require an organization's attention. As you engage yourself in the following chapters, please keep in mind that you should use this book to take your organization to a new level of GIS capability that incorporates an enterprise, sustainable, and scalable solution.

This book is about you as a leader. There are people in this world who could never bring themselves to build something larger and more lasting than themselves, and I am hoping that you are not one of them.

The following nine chapters of this book will show you how to build and manage the implementation of an enduring local government GIS:

1. *Chapter 2*: Strategic Planning
2. *Chapter 3*: The Formula for Success
3. *Chapter 4*: Challenges, Barriers, and Pitfalls
4. *Chapter 5*: Developing a Vision, Goals, and Objectives
5. *Chapter 6*: Governance
6. *Chapter 7*: GIS Training, Education, and Knowledge Transfer
7. *Chapter 8*: Return on Investment
8. *Chapter 9*: How to Sell GIS to Local Government
9. *Chapter 10*: Conclusions

To begin, I have provided a brief summary of each chapter as follows.

1.2 Strategic Planning

Chapter 2 details the strategic planning process that an organization must undertake prior to GIS implementation. Additionally, this chapter discusses the different perspectives on GIS and emphasizes the importance of understanding your audience. Three phases and seven detailed project steps compose the strategic planning process, and the chapter is broken down accordingly.

1.3 The Formula for Success

Chapter 3 is about wrapping your arms around the critical components of local government GIS. Ask yourself where, on a scale of 1 to 10, would your organization fall with regards to its managing and maintenance of a sustainable enterprise GIS? What about your personal understanding of the building blocks of GIS? Could you describe all of the criteria that are required to build a GIS foundation? Can you see the entire GIS landscape? If I asked you to grade yourself on these scores and then explain why you graded yourself that way, could you?

Chapter 3 offers a practical and repeatable strategy for GIS success. It describes the formula for success and documents how you should use it. This entails understanding the essential ingredients that are required to plan, maintain, design, implement, and manage an enduring enterprise GIS. Most importantly, Chapter 3 includes a list of defined GIS terminology and explains how to benchmark your organization according to these definitions.

The heart of Chapter 3 is the formula for success that offers a systematic methodology for examining and benchmarking your GIS.

1.4 Challenges, Barriers, and Pitfalls

There are many obstacles to successful GIS implementation, and unfortunately, the local government landscape is riddled with false starts, poorly planned implementations, and glorified mapping systems. Chapter 4 documents the reason for failure and possible remedies. This chapter's premise is that active management can understand and overcome the challenges to implementation. To support this premise, Chapter 4 discusses five GIS strategic planning components, the planning challenges, and the barriers and pitfalls that can prevent a successful implementation.

It discusses how to sustain the process over multiple years and overcome barriers in the way of change. It discusses the pathways to change and the ways of improving organizational effectiveness and efficiency and lays out the organizational approaches, management processes, and leadership actions that are required for the GIS to become an indispensable part of an organization.

1.5 Developing a Vision, Goals, and Objectives

In the middle ages, English–Welsh archers, or longbowmen, would aim high to hit their target. Over distance, their arrows dropped slightly *en route* to their mark.

Chapter 5 is about aiming high so that you can consistently hit your mark by formalizing goals and objectives. Developing a vision, goals, and objectives will tremendously influence the success of a GIS initiative. In Chapter 5, a detailed seven-step methodology establishes a template for defining your organization's vision and setting goals that support this vision. It describes, in sequential order, how to perform a situation awareness and needs assessment and conduct both Blue Sky sessions and strengths, opportunities, weaknesses, and threats analyses with stakeholders. It addresses how to build consensus and *align* your organization's overall mission with the GIS goals and objectives.

Most importantly, it details the factors that are crucial for building an enterprise GIS vision statement, which includes (a) governance, (b) data and databases, (c) procedures and workflow, (d) GIS software, (e) GIS training and education, and (f) infrastructure. The chapter also explains how to develop performance measures that are related to the stated objectives of an organization.

1.6 Governance

Adding up all the components and subcomponents of GIS technology would make your head spin. GIS has more moving parts than any other technology. It is unique.

Chapter 6 combines theory with real-world experience to offer guidance on the process of managing GIS implementation. It discusses the new set of management challenges that GIS introduces and details what the landscape would look like if local government did not invest in GIS. Through five key components, this chapter introduces a new way to think about GIS technology.

Chapter 6 explains how misguided governance differs from good governance and supports these explanations with a checklist of requirements for a sound governance strategy. It details three different governance models and thus affords you the opportunity, not only to contextualize your organization's current governance model, but also to manage an enterprise GIS according to one of these governance models. It discusses the trend toward regionalization, or a shared-services GIS model, and the role of functional GIS teams, dual accountability, and subject matter experts. It discusses the maturation of GIS and the historic constraints to adoption, including the recent move to *corporate- and cloud-based* style of implementation. Chapter 6 evaluates the best business practices for GIS governance and ends with a list of the important characteristics of a GIS coordinator.

1.7 GIS Training, Education, and Knowledge Transfer

The three important ingredients required to sustain an enterprise GIS are (1) training, (2) education, and (3) knowledge management or knowledge transfer. Chapter 7 discusses a systemic GIS instruction, how to teach GIS skills, and how to create an environment of collaboration whereby the local government can transfer knowledge from one part of the organization to the other. Any organization that builds an excellent process for educating, managing, and transferring knowledge has greater opportunity to build and maintain a value-laden enterprise GIS.

This chapter shows how to develop a training plan for the immediate and future needs of an organization. It discusses the changing nature of GIS technologies and the cycles of GIS design, learning, and use. Chapter 7 shows ways of understanding user needs and, most importantly, building capacity and succession planning.

1.8 Return on Investment

Measuring the success of people and programs is not easy. Even if we know that value may be delivered and experienced through GIS technology, how can we demonstrate this to others?

There are many ways to quantify and qualify the benefits of GIS technology, and Chapter 8 shows how to present the value of GIS technology to local government organizations. It discusses the trend toward a scorecard approach to monitoring government operations, the value of a cost–benefit analysis, the return on investment (RoI) analysis, and the value proposition that is used to explain the benefits of GIS. This chapter introduces 19 RoI categories that are used to explain, document, and detail the real-world examples of how GIS benefits government organizations.

Chapter 8 discusses how GIS technology can save lives, inform and notify the public, prevent local governments from being fined, improve efficiency in virtually all departments, eliminate duplication, and predict events and infrastructure failures. It explains the link between a value proposition and performance indicators.

1.9 How to Sell GIS to Local Government

Do not accept defeat. You can overcome the obstacles to successfully selling GIS to your organization. This process is like comedy: it is all about timing and a great punch line.

Chapter 9 offers a new way of looking at GIS. It is about what GIS can do not only for a local government organization but also for the public and society as a whole. This chapter elaborates on the art of *value propositioning* and explains how to sell GIS technology to elected officials, management, coworkers, and the public. It explains the benefits, both tangible and intangible, of GIS technology and shows you how to do the same.

1.10 Conclusions

Chapter 10 summarizes the entire book. This chapter offers a succinct articulation of the book's message and, most importantly, rehashes what this book contributes to the field of local government GIS.

This chapter simplifies the formula for success, reminds the readers what government professionals are looking for, and offers predictions for the future of GIS technology. Most importantly, Chapter 10 distills the ways that the GIS can be implemented as a truly enterprise and successful solution.

2

Strategic Planning

Give me six hours to chop down a tree and I will spend the first four sharpening the axe.

Abraham Lincoln

2.1 Introduction to Strategic Planning

As they say, failing to plan is planning to fail or, in the words of John F. Kennedy, "Efforts and courage are not enough without purpose and direction." The GIS strategic planning process has changed significantly since I began my career in the industry. Twenty-five years ago, an organization would simply hire a financial advisor to develop a geographic information system (GIS) plan. By the 1990s, the process had somewhat matured. It became a standard for a local government organization to request a consultant to develop a GIS implementation plan. As GIS software and information technology (IT) infrastructure evolved, and the number of GIS users within local government grew, we witnessed a corresponding change in the complexities of the planning process. Somewhere along the way, the words *enterprise* and *strategic* were added.

- *1980–1990*: GIS plan
- *1990–2000*: GIS implementation plan
- *2000–2010*: GIS strategic enterprise plan
- *2010–2020*: GIS strategic enterprise and sustainable business plan

At present, most requests for proposals from towns, cities, and counties ask for a GIS strategic enterprise and a sustainable business plan. One significant change in the past few years has been the requirement of consultants to formulate an optimum *governance* model. This is a new phenomenon. It illustrates how far the GIS has come from a single department to an enterprise solution and, with it, the complexities. All of

9

this certainly qualifies the changing nature of GIS and the GIS planning process.

The process of crafting a *GIS strategic implementation plan (GIS SIP)* includes three basic phases: (1) GIS needs assessment, (2) conceptual system design (CSD), and (3) GIS implementation plan.

2.2 A Proposed Outline for the Scope of Work

It should go without saying that the primary goal of the GIS SIP is to dramatically improve the sustainability, endurance, and enterprise of an organization's GIS solution. But we are playing the obvious game here.

Before we continue, these are a few words to the wise: the scope of the project should not be too simplistic or overelaborate. Disappointment will result from either of these miscalculations. To determine a realistic and successful scope for GIS implementation, an organization should work through the three phases and seven steps that are listed as follows. Lastly, all of the terms that are deployed in the following list are defined in Figure 3.2. While reading through this and other chapters, reference these terms when necessary. The three phases of a GIS SIP include the following:

1. *Phase I: Needs assessment*

   Conduct an objective review of the organization's current GIS capabilities and resources. Identify the GIS data requirements, the functionality, and the ways that the GIS affects workflow in individual departments.

2. *Phase II: System design*

   Provide recommendations on the appropriate hardware, software, staff skills, and governance structures that will support the workflows that are identified in the needs assessment.

3. *Phase III: Implementation plan*

   Develop a thorough implementation plan that details how the system design will meet the goals that are identified in the needs assessment. This plan should detail the timetable and cost schedule that are necessary for an organization to meet these goals.

Figure 2.1 graphically depicts the three phases and seven steps of strategic GIS planning.

The following seven steps break down the ways that phases I, II, and III should be carried out.

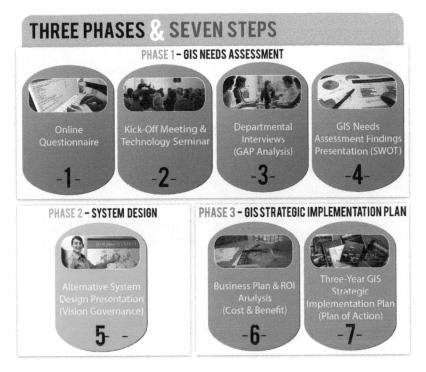

FIGURE 2.1
Three phases and seven steps.

2.2.1 Phase I: Needs Assessment

2.2.1.1 Step 1: Online Questionnaire

Initiate the comprehensive GIS assessment by deploying a custom *online questionnaire* to be distributed to all of the stakeholders in an organization. The questionnaire should be tailored to the specific needs of an organization and should pertain to the organization's existing GIS resources, activities, and workflow. Questionnaires should be delivered to all organizational divisions one to two weeks before conducting *departmental interviews*. The information gathered in step 1, supplemented by an in-depth interview with the GIS coordinator and the GIS staff, is the foundation of the formula for success. Chapter 3 details how to concisely present all of these data.

It is key that the online questionnaire interrogates the following situational points, as they already exist in an organization:

- GIS governance
- GIS digital data and databases

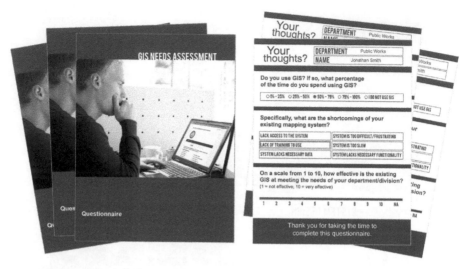

FIGURE 2.2
Online questionnaire.

- Procedures, workflow, and integration
- GIS software
- GIS training, education, and knowledge transfer
- Infrastructure

The natures of these points will be discussed in depth in Chapter 3. Figure 2.2 is an example of an online questionnaire.

2.2.1.2 Step 2: Kick-Off Meeting and Technology Workshop

Use your knowledge of the industry to hold an on-site *kick-off meeting and GIS technology workshop* with all of the potential GIS users within the organization. The combined *kick-off/workshop* should last roughly an hour and a half and provide opportunities for the staff to develop an early, working relationship with GIS technology. This meeting should explain the nature of the needs assessment and GIS SIP to the staff, as well as the scope of the implementation process and the benefits. Figure 2.3 shows an example of an outline of a technology workshop.

2.2.1.3 Step 3: Departmental Interviews

Conduct interviews with all organizational divisions, including all current GIS technology users and all nonusing GIS divisions. Document each stakeholder's and nonstakeholder's roles within the organization and their potential relationship to GIS. It is important during this task to describe the existing business

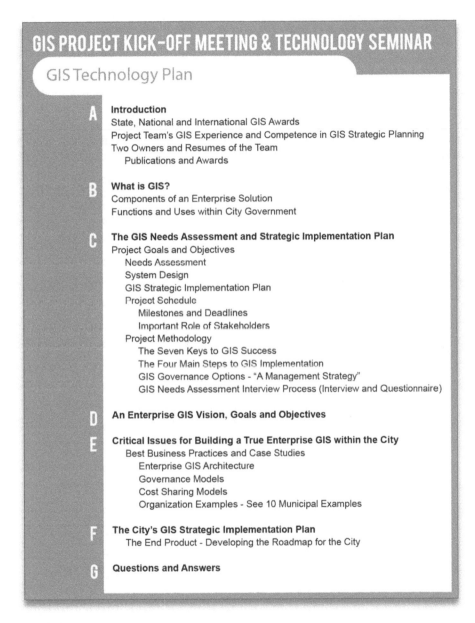

GIS PROJECT KICK-OFF MEETING & TECHNOLOGY SEMINAR

GIS Technology Plan

A **Introduction**
State, National and International GIS Awards
Project Team's GIS Experience and Competence in GIS Strategic Planning
Two Owners and Resumes of the Team
  Publications and Awards

B **What is GIS?**
Components of an Enterprise Solution
Functions and Uses within City Government

C **The GIS Needs Assessment and Strategic Implementation Plan**
Project Goals and Objectives
  Needs Assessment
  System Design
  GIS Strategic Implementation Plan
  Project Schedule
    Milestones and Deadlines
    Important Role of Stakeholders
  Project Methodology
    The Seven Keys to GIS Success
    The Four Main Steps to GIS Implementation
    GIS Governance Options - "A Management Strategy"
    GIS Needs Assessment Interview Process (Interview and Questionnaire)

D **An Enterprise GIS Vision, Goals and Objectives**

E **Critical Issues for Building a True Enterprise GIS within the City**
  Best Business Practices and Case Studies
    Enterprise GIS Architecture
    Governance Models
    Cost Sharing Models
    Organization Examples - See 10 Municipal Examples

F **The City's GIS Strategic Implementation Plan**
  The End Product - Developing the Roadmap for the City

G **Questions and Answers**

FIGURE 2.3
Technology workshop outline.

processes in exhaustive detail so that an organization can storyboard the future application of geospatial technology in relevant and accessible terms.

Perform a *business process analysis (BPA)* for each departmental division in order to define and develop the business activities and GIS workflows that are related to the proposed implementation. The BPA serves as the chief mechanism for documenting the organizational benefit of GIS technology within each department. The needs identified in the BPA are translated, both formally and informally, into the needs assessment.

Departmental buy-in is crucial to the implementation process. Accordingly, each department should have the opportunity to review their assessment for completion and accuracy. Once finalized with the departments, the needs assessment can be used to create a *summary of goals and objectives* that identifies the needs, priorities, costs, and achievability of implementation on a division-by-division basis. This ultimately facilitates the total quantification of organizational priorities.

Note that during this step, mandated obligations may come into play. Mandated obligations refer to any element of the process that must be implemented due to a law, or directive—regardless of the valve to the organization.

STATEMENT OF FINDINGS			
Each department participated	The majority of people do not edit data layers - custodianship and enterprise issues	The majority of people say they use the enterprise GIS	The majority of people have a knowledge of what GIS can do
Most people believed that GIS can benefit them	Most people are not sure about clear lines of responsibility regarding data creation/ maintenance	The majority of city staff use Dakota County web application for GIS use	The majority of people spend 1-25% of their time using GIS
Most people believe that GIS will - Improve data accuracy Improve efficiency Improve information processing	The most important "technical needs" are - Data Standards Digital Submissions Meta Data Data Documentation	The majority of people believe GIS is difficult to use	Most people have received training
Most people see data viewing as the primary GIS activity	Most people consider the areas of "mapping and access to GIS data to be areas of high priority"	Most people do not use GIS on a daily basis	The most important "logistical need" is training and education
Most people believe that the existing GIS is effective at meeting the needs of their department	Key Data Layers are incomplete and inaccurate	Most people use map browsers	Most people believe that GIS training and education is an important component of a true enterprise solution
The most important "strategic need" is departmental goals and objectives		The most important "tactical need" is new uses of GIS	Most people desire more GIS knowledge
The majority of people go to another person in their department for their GIS/ mapping needs		Some users are frustrated by software crashing or slow access to data	Most people would use online training services
		A number of people expressed an interest in mobile GIS	Most people agree that any successful GIS would include - training, education and knowledge transfer

FIGURE 2.4
GIS needs assessment findings.

2.2.1.4 *Step 4: Present the Findings of the GIS Needs Assessment*

Evaluate all of the information that is gathered in steps 1 and 3. The question-naires and interviews should be summarized in a *GIS needs assessment report* that details key implementation issues and departmental needs. Present this report to the organization in a briefing that includes an executive-level sum-mary, a PowerPoint presentation, and the opportunity for discussion and feedback. A detailed report will comprise findings that include the potential applications, the necessary data, the required resources, and the identified workflows. This report should undergo one review and edit cycle with the full GIS project team. Figure 2.4 shows an example of the findings of a GIS needs assessment.

2.2.2 Phase II: Alternative Conceptual System Design

2.2.2.1 *Step 5: Developing the CSD*

The CSD is the first step in the actual development of an organization's GIS. This is the blueprint for the implementation of an enterprise GIS and is based on a review of the data that are gathered in steps 1–4. Alternative CSD is a characteristic of a good GIS SIP.

In order to develop a CSD, an organization should compartmentalize the following points with regards to the findings of their needs assessment:

- GIS governance

  A complete understanding of an organization's existing governance model is required. A diagrammatic illustration of the lines of com-munication and a determination of the actual governance model that is used (centralized, decentralized, or hybrid) should be given. A new proposed governance strategy or model should be detailed, documented, and most importantly, explained.

- GIS digital data and databases

  A digital data assessment of either every data layer or just the foun-dation layers of an enterprise GIS, including parcels, address points, street centerline, and aerial photography should be given. An enter-prise database readiness assessment should detail the quantity, quantity, and completeness of the existing digital data layers.

- Procedures, workflow, and integration

  All the procedures, workflows, and opportunities for GIS integra-tion should be detailed and considered. New GIS-centric procedures and protocols need to be documented in the CSD.

- GIS software

  The CSD should include a GIS software plan that considers all existing and future GIS software solutions and users. Each department should be evaluated and assigned a type of GIS user, including tier 1, 2, 3, and 4. (Figure 7.2; Four tiers of GIS users.)

- GIS training, education, and knowledge transfer

  A multiyear and sustainable training, education, and knowledge transfer plan needs to be created as part of the CSD.

- Infrastructure

  A design specification needs to be created to support the enterprise and sustainable GIS. An optimum and sustainable system architecture design and a plan to upgrade and deploy new IT infrastructure need to be created.

2.2.3 Phase III: Final Implementation Plan

2.2.3.1 Step 6: Business Plan

The *GIS business plan* should incorporate the background information that is gathered during phase I: needs assessment and phase II: system design. It should focus on detailing and documenting the value of the GIS investment for all the stakeholders within the organization, including each department, the enterprise, and the public. The business plan should include the use of the 19 RoI categories that are detailed in Chapter 8. It is important to align the vision and goals and objectives of the organization with the goals and tasks of the proposed enterprise GIS. The three alternatives for developing a business plan include the following:

1. Cost–benefit analysis
2. RoI analysis
3. Value proposition

Whichever level of analysis you choose should outline the project's goals and justify them via an explanation of their attainability. It should outline a plan for attaining the goals and describe the performance measures that will be used to gauge whether or not these goals have been met.

Ultimately, you are charged with describing and justifying how investment in the GIS can benefit the organization and its citizens. You should be able to answer the question, why should the organization invest in the GIS? Figure 2.5 shows the outline of a business plan.

GIS BUSINESS PLAN OUTLINE

INTRODUCTION

PROJECT OUTLINE AND STRATEGIC VISION, GOALS AND OBJECTIVES

RECOMMENDED GIS IMPLEMENTATION ALTERNATIVIES

MULTIYEAR BUDGET

    Year One

    Year Two

    Year Three

    Ongoing Costs

    Funding Issues

        Capital Costs

        Operational Costs

        Enterprise Funds

BENEFITS AND BUSINESS JUSTIFICATION TO THE ORGANIZATION

    Tangible Benefits

        16 ROI Categories

            Value Proposition

            Cost-Benefit Analysis

            ROI Analysis

    Intangible Benefits

MEASURING SUCCESS

    Key Performance Indicators (KPI)

    Quantifiable Performance Measures

RISKS

REGIONALIZATION AND SHARED COST OPTIONS

SUCCESSION PLANNING ISSUES

FUTURE TECHNOLOGY

FIGURE 2.5
GIS business plan.

2.2.3.2 *Step 7: Final Implementation Plan and Presentation (Plan of Action)*

Develop a GIS SIP that recommends changes to the organization's GIS program, supported by the information that is gathered during the assessment. A multiyear-phased GIS implementation plan should include the following:

- A step-by-step plan of action
- Multiple and discrete phasing
- Funding options and alternatives
- Short- and long-term solutions
- Implementation costs

An hour-long presentation that explains the preceding points should be given. A question-and-answer period should conclude the session, and copies of an executive summary should be provided to the stakeholders who are present. Figure 2.6 illustrates the difference between a GIS SIP and a GIS Checkup. My point here is to illustrate that you can perform a rapid assessment of your organizations GIS. This assessment will guide your future directions. Figure 2.7 outlines the GIS SIP.

The three-phased strategic planning methodology detailed in this chapter is the most succinct and straightforward explanation of what should be included in the planning methodology. After the completion of a GIS SIP, consider the actual implementation process itself and the costs that are associated with implementation. Figure 2.8 graphically illustrates a four step simple process of looking at the entire GIS implementation project. Figure 2.9 considers five budgetary phases of GIS implementation. You should consider how many phases or steps are in the entire implementation process and plan accordingly. My point in this book is that I believe that the GIS SIP should use three phases and seven steps. After that, use a method you feel appropriate for presenting the entire implementation process and all the costs that are associated with each component.

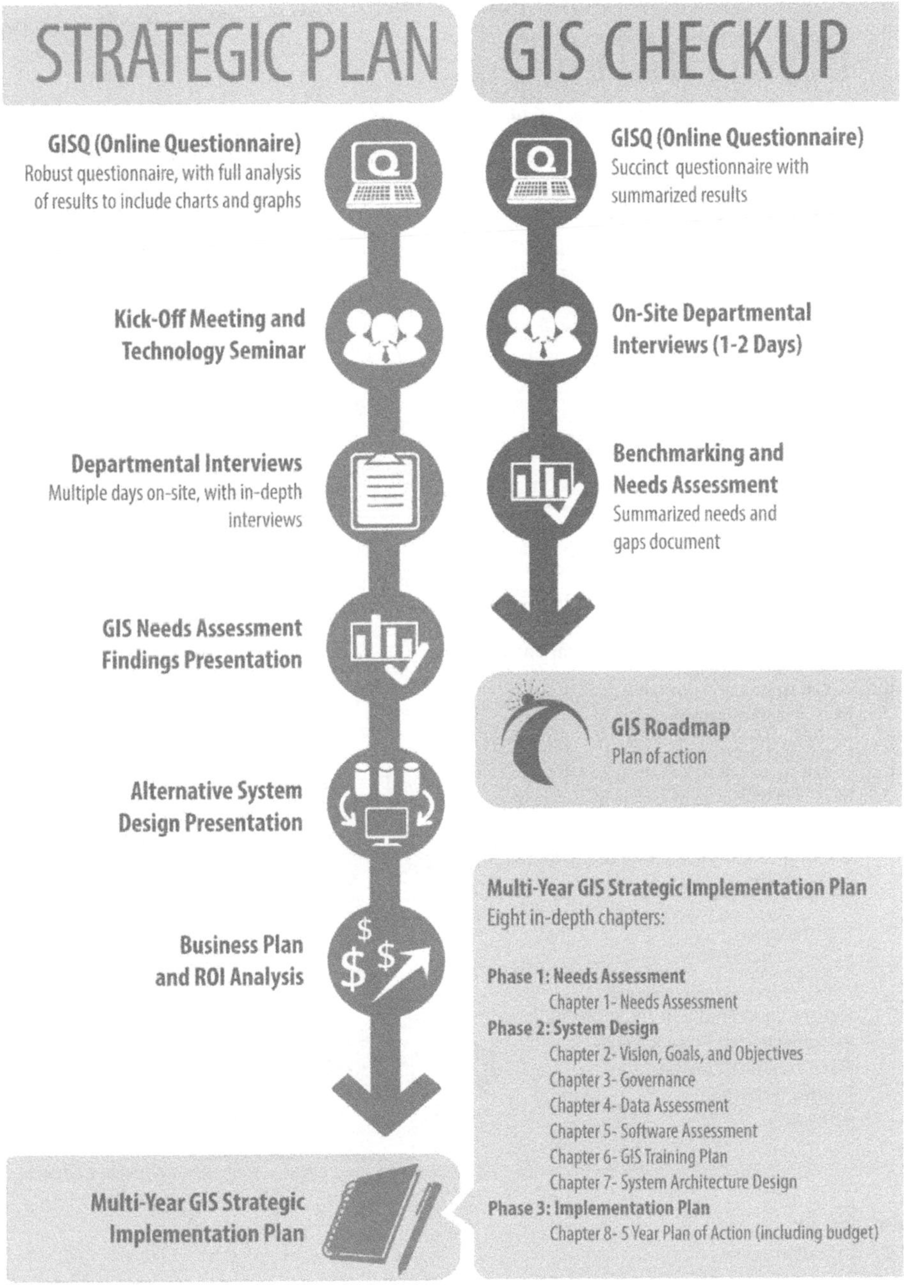

STRATEGIC PLAN

GIS CHECKUP

GISQ (Online Questionnaire)
Robust questionnaire, with full analysis
of results to include charts and graphs

GISQ (Online Questionnaire)
Succinct questionnaire with
summarized results

**Kick-Off Meeting and
Technology Seminar**

**On-Site Departmental
Interviews (1-2 Days)**

Departmental Interviews
Multiple days on-site, with in-depth
interviews

**Benchmarking and
Needs Assessment**
Summarized needs and
gaps document

**GIS Needs Assessment
Findings Presentation**

GIS Roadmap
Plan of action

**Alternative System
Design Presentation**

Multi-Year GIS Strategic Implementation Plan
Eight in-depth chapters:

**Business Plan
and ROI Analysis**

Phase 1: Needs Assessment
Chapter 1- Needs Assessment
Phase 2: System Design
Chapter 2- Vision, Goals, and Objectives
Chapter 3- Governance
Chapter 4- Data Assessment
Chapter 5- Software Assessment
Chapter 6- GIS Training Plan
Chapter 7- System Architecture Design

**Multi-Year GIS Strategic
Implementation Plan**

Phase 3: Implementation Plan
Chapter 8- 5 Year Plan of Action (including budget)

FIGURE 2.6
GIS strategic plan versus GIS checkup.

GIS STRATEGIC
IMPLEMENTATION PLAN

TABLE OF CONTENTS

Project Overview
Online Questionnaire Results
Departmental Needs Assessment
  Airport
  Attorney's Office
  Board of Elections
  Building Permits and Inspections
  Central Services
  Computer Information Services (CIS)
  Emergency Management
  Finance
  Fire and Emergency Services
  Housing and Community Development
  Human Resources
  Leisure Services
  Manager's Office
  Planning
  Police
  Public Information Office
  Public Utilities
  Solid Waste
  Tax Assessor's Office
  Transportation and Public Works
Existing and Future Departmental Software Users
  Generic Software Descriptions
  Software Departmental Matrix
Benchmarking Analysis
  Governance
  Data and Databases
  Procedures and Workflow
  GIS Software
  Training and Education
  Infrastructure
General Findings and Recommendations
  Governance
  Data and Databases
  Procedures and Workflow
  GIS Software
  Training and Education
  Infrastructure
  Infrastructure Recommendations
Return on Investment (ROI) Analysis
Recommended Project Steps
Recommended Budget
Project Schedule
Appendix

FIGURE 2.7
GIS SIP outline.

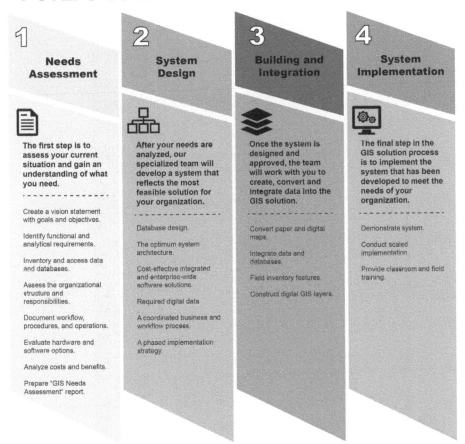

4 STEPS TO IMPLEMENTATION ▸

1 Needs Assessment

The first step is to assess your current situation and gain an understanding of what you need.

- - - - - - - - - - - - - -

Create a vision statement with goals and objectives.

Identify functional and analytical requirements.

Inventory and access data and databases.

Assess the organizational structure and responsibilities.

Document workflow, procedures, and operations.

Evaluate hardware and software options.

Analyze costs and benefits.

Prepare "GIS Needs Assessment" report.

2 System Design

After your needs are analyzed, our specialized team will develop a system that reflects the most feasible solution for your organization.

- - - - - - - - - - - - - -

Database design.

The optimum system architecture.

Cost-effective integrated and enterprise-wide software solutions.

Required digital data

A coordinated business and workflow process.

A phased implementation strategy.

3 Building and Integration

Once the system is designed and approved, the team will work with you to create, convert and integrate data into the GIS solution.

- - - - - - - - - - - - - -

Convert paper and digital maps.

Integrate data and databases.

Field inventory features.

Construct digital GIS layers.

4 System Implementation

The final step in the GIS solution process is to implement the system that has been developed to meet the needs of your organization.

- - - - - - - - - - - - - -

Demonstrate system.

Conduct scaled implementation.

Provide classroom and field training.

FIGURE 2.8
Four steps to GIS implementation.

Population	Geographic Technologies Group (Representative Clients)	FIVE PHASES OF IMPLEMENTATION					TOTAL COST
		Phase I: Strategy and Planning	Phase II: Analysis and Design	Phase III: Development	Phase IV: Deployment	Phase V: Production and Operation	
125,000	City in Ontario, Canada	$30,000	$57,000	$325,000	$165,000	$75,000	$652,000
166,179	City in Ohio	$25,000	$67,000	$295,000	$215,000	$110,000	$712,000
433,575	City in Virginia	$30,000	$70,000	$750,000	$350,000	$320,000	$1,520,000
90,329	City in Florida	$15,000	$62,000	$300,000	$150,000	$60,000	$587,000
48,080	City in Missouri	$12,500	$38,000	$300,000	$150,000	$50,000	$550,500
44,756	City in Florida	$10,000	$58,000	$35,000	$100,000	$95,000	$298,000
39,728	City in Maryland	$18,000	$55,000	$300,000	$130,000	$75,000	$578,000
73,000	City in Alabama	$12,500	$58,000	$390,000	$100,000	$75,000	$635,000
338,774 | 227,834	City & County in North Carolina	$32,000	$68,000	$1,400,000	$450,000	$300,000	$2,250,000
246,681	County in California	$28,000	$72,000	$1,200,000	$450,000	$200,000	$1,950,000
3,010,759	County in California	$70,000	$128,000	$1,500,000	$450,000	$400,000	$2,548,000
190,557	County in Arizona	$25,000	$58,000	$1,200,000	$400,000	$200,000	$1,883,000
3,836	City in Alaska	$15,000	$38,000	$225,000	$100,000	$30,000	$408,000
75,254	City in Illinois	$12,500	$38,000	$250,000	$150,000	$100,000	$550,500
86,474	County in Maryland	$17,500	$48,000	$300,000	$300,000	$100,000	$765,500
33,698	County in Wyoming	$20,000	$54,000	$240,000	$180,000	$75,000	$569,000
84,644	County in Maryland	$25,000	$48,000	$275,000	$125,000	$100,000	$576,000
AVERAGE		$23,411.76	$59,823.53	$546,176.47	$233,235.29	$139,117.65	$1,001,911.76

FIGURE 2.9
GIS implementation costs.

3

The Formula for Success

The most important single ingredient in the formula for success is knowing how to get along with people.

Theodore Roosevelt

3.1 Introduction to the Formula for Success

The bedrock of my formula is a thorough understanding of one's audience and one's organization. It is the people in an organization who will determine its success or failure. Not the technology. Not the funding. The people. It goes without saying that the *human factor* is the single most important component in GIS implementation.

Local government professionals bring a diversity of perspectives to the table. Every one of these professionals carries a different technical, social, or managerial DNA. More often than not, each professional has a completely different *take* on the adoption and management of GIS technology. Understanding these perspectives will significantly influence the success of your organization and its relationship with geospatial technology.

If people are the absolute key to success, a close second is the detailed planning process for implementing and maintaining a successful GIS enterprise. This planning process has many moving parts: (a) six strategic challenges, (b) a phased approach to implementation, and (c) a formula that allows your organization to see itself holistically and benchmark itself against other organizations.

Figure 3.1 provides six graphics that are rolled into one. It illustrates a step-by-step outline to develop an enterprise GIS strategic, sustainable, and business-oriented implementation plan. The linear phases include the following:

- Understanding the history of GIS adoption and the barriers to success
- Recognizing organizational issues and challenges
- Survey, teach, listen, explain, and sell

FIGURE 3.1
Heuristic view of GIS strategic planning.

- Alternatives, options, and benefits
- Business alignment and value proposition
- Phased multiyear implementation plan

3.2 Different Perspectives on GIS

What are people thinking, and why? Though we can never know for sure, a sincere understanding of personalities and backgrounds is as important to success as any technology. When dealing with people in your organization, try to remember the words that were uttered by that famous philosopher Popeye the Sailor Man: "I yam what I yam, and that's all what I yam."

I would explain GIS to upper management in a far different way from how I would explain it to a class of fourth graders. Presenting geospatial technology in Wyoming or Alaska is different from presenting it in the Bahamas, Abu Dhabi, or Canada. Your message must resonate with your audience, just as the technologies resonate with the landscape.

It is important to understand that adopting GIS as a new standard is one thing, but implementing the technology is a completely different *kettle of fish*. It may require new practices in the workplace and ongoing activities that are related to implementation. Some would say that, by its multifaceted nature, GIS implementation is a problematic, or at least problem-prone, process. Thus, understanding the perspectives of the decision makers in a local government organization is crucial to a successful implementation.

We can agree that government officials want one thing from any project: measurable results. While everyone in the organization may have different day-to-day concerns, this is a shared ultimate goal. Here is a breakdown of the areas of concern for various levels of government officials:

- *Elected officials* may have little concern for GIS functionality, but they do want to know how the GIS can deliver a return on taxpayers' money, how investing in GIS technology can make their communities a safer place to live, and how GIS can increase government transparency. How GIS aligns with the organization overall vision is critical to elected ofiicials.

- *Government administrators* must not only oversee the adoption and management of GIS but also understand ways to invest resources that will benefit all of the stakeholders. Clear lines of responsibility, accountability, alignment, and measurable results are their chief concerns.

- *Information technology (IT) directors and GIS coordinators* concentrate on the challenges of *bringing* the technology to the users: designing GIS architecture, managing bandwidth demands, budgeting for software acquisition, and training personnel. Let us not forget the actual decision support, analysis, and predictive modeling routines that a GIS coordinator deals with on a day-to-day basis.

3.3 A Layman's Discussion of Campbell and Masser's Book *GIS and Organizations*

Authors who have *all-around* exceptional qualities and produce *page-turners* like Heather Campbell and Ian Masser have offered us an incredible insight into the psychology of local government professionals. After reading Campbell and Masser's book, *GIS and Organizations* (1995), I was able to categorize and describe my own observations about the local government perspectives on GIS and relate them to the process of successful GIS implementation using the approach that Campbell and Masser articulate.

The following section represents information about the different perspectives on technology. I like to think of it as a layperson's adaptation of Campbell and Masser's perspectives on implementation that I have tied to real-world examples. Campbell and Masser describe three groups of GIS perspectives: (1) *technological determinism*, (2) *managerial rationalism*, and (3) *social interactionism*. The following section explains the ways that I have seen these perspectives interact with GIS, first, by outlining their goals and mind-set and second, by explaining the pitfalls of allowing a homogenous group of said individuals to implement GIS.

3.3.1 *Technological Determinism*: The IT Guru or Tech Geek Perspective

The *technical expert* or *IT guru* is a character type who is generally found in the IT department of local government. He or she has an attitude that says, "The GIS is so easily understood," and he or she tends to regard systemic implementation as a purely technical process.

Throughout my 25 years in the industry, I have never witnessed an IT guru regard GIS implementation as anything other than a technological problem and one that can be solved fairly quickly. These individuals are, as Campbell articulates, "guided purely by the inherent value of the technological innovation." The technical guru in your organization is purely focused on the technological advantages of geospatial technology.

Generally speaking, this group has a specific mind-set that values the following:

- New and innovative technology as the most important factor.
- Technological advantages are regarded as valuable, in and of themselves.
- Implementation is a technological issue, much like a programming problem.
- GIS technology can be deployed to solve an operational problem.
- The barrier to GIS implementation will be the ignorance of the *users*. It will never be the technology.
- They will support immediate utilization and new working practices (mostly for other departments).
- They believe that GIS implementation should reside in the hands of the technical experts and that the individuals who possess these skills are essential to the initiative.
- Undisciplined or ill-educated staff will be a barrier to success.
- They generally lack a picture of the broader enterprise.
- Personal goals are extremely important to these individuals.

As a result of their overwhelming understanding of technology as an end unto itself, the tech gurus often overlook the human factor that is crucial to the successful implementation of GIS. Thus, results like those in the list above often manifest when these individuals manage implementation. There are many examples within local government where it gurus have stiffled GIS growth.

3.3.2 *Managerial Rationalism*: The Local Government Management Perspective

It has been my pleasure to work with many senior management professionals over the years. This group understands their core position in successful government operations and views GIS implementation through the lens of management principles and techniques. They understand the language of management and use it to describe the GIS implementation process as they see it: an intersection of new technology and a strategy that is developed by a proven process within their organization.

Do not expect senior management to implement GIS as a result of its potential value, but rather as a tool that is empirically proven to bring their managerial strategy to life in an organization. Management officials bring the following mind-set to the table:

- Management techniques will drive their organization's implementation of a GIS strategic plan.

- Implementation will *only* occur if management embrace the solution.
- Poor management will prevent GIS from being successful. This is an Achilles heel for GIS implementation.
- Managers act rationally, weighing marginal costs and marginal benefits.
- Their personal goals are secondary to the goals of the organization, or rather, their personal goal *is* often the success of the organization as a whole.
- Strategic activities and good decision making by management will drive home a successful GIS implementation.

3.3.3 *Social Interactionism*: The Local Government Individual Perspective

The local government individual's perspective on GIS tends to embody the idea behind Campbell and Masser's category of social interactionism. I use the term *local government official* to describe the members of local government who do not fit into the technical guru or management categories. This category comprises the majority of staff in local government. They believe that successful GIS implementation is based on their support for GIS, in conjunction with effective social interaction and cooperation.

Change management is the art of responsibly managing change within an organization. It is crucial to the perspective of local government individuals. Additionally, these folks understand that cooperation plays a significant role in implementing GIS. These are factors that cannot be ignored when dealing with local government officials. Their mind-set tends to be built around the following:

- The style of GIS implementation is influenced by the individuals within an organization more than anything else. These individuals are metered not only by their formal qualifications but also in their ability to adapt and change.
- GIS adoption and implementation can be influenced by social trends, particularly among this group.
- This group focuses heavily on the symbolic status of power.
- The environment or culture among an organization's employees factors heavily into GIS implementation.
- Organizational and user acceptance of GIS technology plays a key role.
- The personnel implications of GIS are important.
- The intangible benefits of GIS are important.

3.4 The Truth of the Matter

Understanding the various perspectives on the adoption and implementation of GIS is a critical ingredient to the project's success. There will always be the technical, managerial, and individual aspect of GIS within local government, and understanding the multiplicity and nuance of these perspectives provides a solid foundation for success.

3.5 The Formula for Success—A Checklist

Perhaps the most rewarding part of GIS strategic planning is the ability to effectively and efficiently evaluate the existing GIS conditions of an organization and then, in turn, use this information to develop an understandable action plan for the future.

It was in 2010 when I realized that the *long-drawn-out* process of the GIS *strategic planning* process needed to be organized and streamlined. So, about five years ago, I decided to keep notes on the most important factors that went into being a *great* GIS coordinator or manager. These factors ultimately became a comprehensive checklist of tasks that, if followed and implemented, would make for a very successful and enterprise GIS. Essentially it became my *playbook* for all GIS coordinators. After a few years of roaming around the United States developing GIS plans, I had a list of enough essential tasks that I was forced into incorporating them into six understandable categories. These six categories, supported by the tasks that make up the formula for success, allowed me to rapidly evaluate an organization and (more by luck than planning) benchmark that organization against other organizations with similar characteristics. What I am saying is that if you can *grade* yourself on each of the following tasks, you should be able to understand your strengths and weaknesses, threats, and opportunities. Let us take a look at my foundation for the formula for success. The idea is to grade an organization on how successful it has been at meeting or exceeding the essential goals of a successful enterprise GIS, listed below. Remember that we are presupposing that my categories and tasks are the formula for success or the essential ingredients in building an enterprise GIS. For example, some questions are answerable by a *yes* or a *no*. The first question below is "Do you have a GIS strategic plan?" If the answer is yes, you should award the organization 100%. I tend to deduct 10% points for the age of the plan, one year equals 10%. A GIS strategic plan that is six years old would get a 40% rating as I deducted 60% for each of the six years. Some of the subcategories listed below are actually questions and require a judgment call. For example, "Does the organization embrace regionalization of GIS?" The grade is based on how we define regionalization. Let us look at each essential goal.

3.5.1 Category One: GIS Governance

Category one: GIS governance is described by the following tasks. Figure 3.2 illustrates all of the key governance components of an enterprise, sustainable, and enduring GIS solution in local government.

- *A GIS strategic plan*

  Whether you are hosting a dinner party, organizing a business retreat, or taking your family on vacation, a plan of action is crucial to the success of your endeavor. GIS technology is no different. A sound *GIS strategic implementation plan (GIS SIP)* provides the game plan for an organization's development of a successful relationship with geospatial technology. The GIS SIP integrates the varying levels of organizational concerns into a formalized architecture for the implementation of GIS technology. Let us make that a little clearer with a simple example. We could say that a municipal government organization's overall goal is to improve life for its citizens. The GIS SIP takes this overall *vision* and breaks it down into concrete *goals*.

GOVERNANCE COMPONENTS

- ☑ A GIS Strategic Plan
- ☑ Annual Update to the Strategic Plan
- ☑ A GIS Vision, Goals and Objectives
- ☑ A Formalized Governance Model
- ☑ Job Classifications
- ☑ Enterprise GIS Project Management
- ☑ Coordinated GIS Enterprise
- ☑ GIS Steering Committee
- ☑ GIS Sponsor Team
- ☑ GIS Technical Committee
- ☑ GIS Functional Groups
- ☑ GIS User Groups
- ☑ Regionalization of GIS (MOU and Data Sharing)
- ☑ GIS Policy and Mandates
- ☑ User Sensitivity
- ☑ GIS Collaboration
- ☑ Measure Quality of Service
- ☑ GIS Authority and Clear Lines of Responsibility
- ☑ A GIS Budget or Funding Model
- ☑ Charge Back Model
- ☑ Grants and Funding Initiatives
- ☑ An Annual Detailed GIS Work Plan
- ☑ GIS Coordination Tasks
- ☑ Key Performance Measures or Indicators (KPI)
- ☑ GIS Blog or Newsletter
- ☑ A GIS Culture or Collaboration
- ☑ Alignment with Organization's Visions, Goals and Objectives
- ☑ Service Level Agreement (SLA)
- ☑ Cost Recovery
- ☑ Revenue Generation

FIGURE 3.2
Governance components.

These goals could be something such as a more efficient deployment of parks and recreation department resources or improved response times for municipal emergency services. Both of these goals serve the larger vision of making the citizenry happier and healthier. In this sense, the GIS SIP makes the abstract concrete. Furthermore, the GIS SIP provides a framework for an organization to achieve these goals. During the planning process, an organization identifies the areas in need of improvement and then develops metrics that rate the process of improvement. Drawing from our earlier example, the metric may be *average EMS unit response time*. The GIS SIP then details the tactics and strategies that are necessary for the deployment of GIS technologies and how these tactics and strategies will improve the average response time. Juxtaposed against this presentation of benefits will be a cost projection and overall value analysis.

- *Annual update to the strategic plan*

    The strategic plan should be updated annually. Organizations are organic. Their roles, vision, and functions constantly evolve. The strategic plan should be updated to stay relevant to the organization's vision and the practical aspects of implementation.

- *A GIS vision, goals, and objectives*

    As previously mentioned above, the larger vision (higher quality of life for citizenry) of an organization must be broken down into concrete goals (improved emergency response time). The vision, goals, and objectives of GIS technology must align with the organization's vision and have measurable *objectives*.

- *A formalized governance model*

    The term governance model refers to the constellation of relationships between individuals and departments within an organization. A governance model lays out lines of responsibility and the hierarchy of decision making power within an organization. These lines connect executives, managers, and staff, or more broadly the *stakeholders*. A stakeholder is any individual directly affected by an organization's activities. Formalizing a governance model allows an organization to maximize accountability and efficiency. It designates the tasks each organizational entity must accomplish.

- *Job classifications*

    The various positions within an organization should be classified according to the formalized governance model. These job classifications denote the skill set, financial worth, decision-making power, hierarchical standing, and overall responsibilities of a given position within the organization. Chief executive officer, marketing manager, and recreation worker exemplify standard job classifications. Keep in mind that these job classifications may need adjustment during the GIS implementation process. A recreation worker who formerly signed patrons in at a municipal swimming pool on a clipboard, may, following implementation, enter patron data into a GIS technology application. The increase in technological skills and problem-solving skills may warrant a reevaluation of the recreation worker's job classification.

- *Coordinated GIS enterprise*

    A coordinated GIS enterprise refers to a situation where an organization's GIS governance model allows for a GIS coordinator to oversee and coordinate all GIS projects as if they were part of the enterprise. That is to say, all GIS projects are managed to a lesser or greater extent by a centralized group. Though they interact with other divisions on a daily basis, the GIS coordination's division holds the ultimate responsibility for administering, monitoring, and developing the larger organization's geospatial technologies.

- *GIS steering committee*

    A GIS steering committee is a group that is composed of top-level organizational leaders and GIS specialists. This group often includes all departmental directors of an organization, along with top financial and administrative officers and the GIS coordinator. The steering committee allocates resources for the organization's GIS needs and determines the schedule, priority, and policy issues that are related to implementation. A coherent GIS steering committee is crucial for a smooth implementation process, as it allows direct interfacing between executive decision makers and GIS experts.

- *GIS sponsor team*

    The GIS sponsor team is composed of executive leadership or an executive leader. This person(s) is responsible for championing the GIS cause, resource acquirement, and budgeting that is related to the GIS implementation process. The sponsor team also mediates the relationship between GIS directives and the organization's larger vision.

- *GIS technical committee*

    As the name implies, the GIS technical committee oversees all of the technical challenges of deploying an enterprise GIS. It sets standards for ways that GIS data are gathered, managed, and shared in an organization. Most of what this committee does is related to systems architecture and IT infrastructure.

- *GIS functional groups*

    GIS functional groups are specialized teams within an organization responsible for discussing and overseeing key focus areas, including public safety, land management, administration, and utilities. Functional groups are created usually when the organization is large and complex. These groups essentially divide the task of the GIS steering committee up into management components. They are, by their nature, narrow in focus and require some degree of expertise from their members.

- *GIS user group*

A GIS user group is a cohort of stakeholders who share information and compare experiences with GIS technology for the benefit of all members. A GIS user group is managed by the GIS coordinator and meets frequently, often every month or each quarter.

- *Regionalization of GIS*

Regionalization is a formalized agreement between parties or entities to cooperate. In relation to geospatial technologies, regionalization is the sharing of data, resources, applications, training, and education and more between disparate groups of GIS users in the region seeking to pool their resources and achieve similar goals. Often, memorandums of understanding (MoUs) guide the regionalization of GIS technologies, where multiple organizations, grouped by geography, share data with one another.

- *GIS policy and mandates*

Policies refer to procedural codes of conduct that are ratified and enforced by organizational authorities. These policies are internally imposed and guide everything from data and resource sharing within the GIS initiative, to personal and financial concerns for the organization at large. Mandates are externally imposed (often by larger governmental bodies or organizations) guidelines that an organization must follow.

- *User sensitivity*

User sensitivity refers to the capabilities of a particular GIS technology to fluidly respond to a user's request for information. User sensitivity is an important measure of the relative benefits of implementing GIS technology. User sensitivity can be managed by using questionnaires, one-on-one interviews, GIS user group feedback, and more.

- *GIS collaboration*

GIS collaboration refers to the productive cooperation between individuals and entities facilitated by the implementation of GIS technology. High levels of GIS collaboration let an organization, or organizations, derive maximal benefits from enterprise GIS technologies. It is both a by-product and end goal of geospatial technology.

- *Measure quality of service*

    Measuring quality of service refers to an organization's capacity to gather feedback data about the efficacy of its geospatial technologies. The quality of service can be examined through questionnaires and interviews or metrics that are related to user interface and objective goals.

- *GIS authority and clear lines of responsibility*

    A line of responsibility describes the vertical chain of liability and authority in an organization. In common-sense terms, a line of responsibility formally lays out who is responsible for what and to whom.

- *A GIS budget or funding model*

    A funding model is a methodical and institutionalized approach to building a reliable revenue base to support an organization's core programs and services. In our context, an organization's funding model explains, in formal financial terms, how the geospatial technology initiative will be funded.

- *Chargeback model*

    Instead of paying for an initiative from a centralized GIS budget, in a chargeback model, individual departments pay for the GIS services that they utilize. Instead of a centralized GIS budget that covers all the costs that are associated with organization-wide geospatial technology, each user essentially pays for software, services, and support. A chargeback model can often stifle growth and is not seen as an optimum solution.

- *Grants and funding initiatives*

    A funding initiative allows a government organization to diversify the funding for its GIS initiative. Grants are sums of money that are distributed by governmental entities for specific project-related purposes. A local government organization should review all opportunities for grant funding to support the GIS initiative. Also, many local government organizations have what is called enterprise funds that can be used for a GIS initiative.

- *An annual detailed GIS work plan*

    A work plan proposes the schedule and budgeting for a specific project. It not only offers a step-by-step description of the ways that a plan will be enacted but also projects a timeline and explains how funding will be deployed within the plan's framework. The work plan associated with a GIS initiative should be updated on an annual basis to reflect the evolving needs and priorities of a GIS enterprise organization. Essentially, it lays out a work plan for the GIS team as it relates to the priorities of the GIS steering committee and each department.

- *GIS coordination tasks*

  The GIS coordinator must coordinate and participate in all the GIS *tasks* within the organization. Objective-driven assignments are given to each department or individual within an organization. The GIS coordinator supports and manages each GIS project.

- *Key performance measures or indicators*

  Key performance measures or key performance indicators (KPIs) are organizationally ratified metrics that gauge whether and how specific goals are met by an organization. These objectives, numeric representations of success or failure are crucial when comparing the costs and benefits of the GIS initiative.

- *GIS blog or newsletter*

  A GIS blog or digital newsletter is produced by an organization in order to increase communications around a GIS initiative. It provides transparency and accountability by keeping stakeholders and citizens in the loop through easily accessible media.

- *A GIS culture of collaboration*

  A culture of collaboration refers to an attitude that is expressed by stakeholders in their relationships to one another, as it pertains to an enterprise GIS. It is an unquantifiable web of positive interpersonal interactions that facilitates creative problem-solving and resource sharing among individuals and departments to achieve commonly held goals.

- *Alignment with organization's vision, goals, and objectives*

  The enterprise GIS needs to be aligned with the organization's vision, goals, and objectives; otherwise, it serves no purpose. This is necessary from the ground up. Simplistically, the vision of an organization may be to improve life for its citizenry. Enterprise GIS supports this vision by identifying areas that need improvement and giving decision makers the capacity to set realistic, data-backed goals (such as the improved emergency service response time). These goals would then be broken down into objectives to be measured by KPIs.

- *Service level agreement*

  Service level agreements (SLAs) are formal, legally binding agreements that outline what stakeholders can expect from enterprise GIS. The parameters of an SLA are defined by the KPIs that are relevant to the technologies in question. Essentially, an SLA can be created to document how the GIS group will support each department.

- *Cost recovery*

  Cost recovery is exactly as the name implies. A cost recovery policy within an organization mandates that the organization will recover the costs for the act of responding to citizen and business requests for data. It may include staff and computer time, as well as hardware expenses including thumb drives or compact discs. Essentially, the organization is not charging for data; it is recovering the cost to make it available.

- *Revenue generation*

    Revenue generation is a policy whereby an organization can actually charge for GIS data and services. Essentially, this is when an organization charges a fee above and beyond just cost recovery. The philosophy is that a price can be set for GIS services to essentially pay for the entire cost of implementing and maintaining its GIS program.

Figure 3.3 illustrates the end product after a self-assessment of the existing governance conditions of the organization. This figure graphically illustrates

Governance Components	0%	10%	20%	30%	40%	50%	60%	70%	80%	90%	100%
A Strategic Plan				25							
Annual Update to the Strategic Plan											
A GIS Vision, Goals, and Objectives											
A Formalized Governance Model											
Job Classifications								70			
Enterprise GIS Project Management				30							
Coordinated GIS Enterprise				30							
GIS Steering Committee											
GIS Sponsor Team											
GIS Technical Committee					40						
GIS Functional Groups											
GIS User Group											
Regionalization of GIS (MOU and Data Sharing)						50					
GIS Policy and Mandates				30							
User Sensitivity			15								
GIS Collaboration		10									
Measure Quality of Service				30							
GIS Authority and Clear Lines of Responsibility				30							
A GIS Budget or Funding Model								70			
Charge Back Model				25							
Grants and Fundings Initiatives		10									
An Annual Detailed GIS Work Plan											
GIS Coordination Tasks			15								
Key Performance Measures or Indicators (KPI)											
GIS Blog or Newsletter											
A GIS Culture of Collaboration			15								
Alignment with Organization's Vision, Goals, and Objectives											
Service Level Agreement (SLA)											
Cost Recovery											
Revenue Generation											

Governance Self-Assessment

FIGURE 3.3
Governance self-assessment.

the end graphic after a self evaluation of all governance components required in local government.

3.5.2 Category Two: GIS Digital Data and Databases

Category two: GIS digital data and databases is described by the following tasks. Figure 3.4 illustrates all of the key data and database components of an enterprise, sustainable, and enduring GIS solution in local government.

- *A digital data assessment and review*

  A digital data assessment examines the completion and breadth of an organization's existing data layers. It evaluates the accuracy, completeness, and overall health of the existing digital data layers within an organization. Once the data are assembled, gaps and weaknesses are identified and subsequently improved.

- *Master data list*

  The master data list (MDL) enumerates all of the data sets that an organization needs for enterprise GIS implementation. The various data sets should be detailed by type and source, and assessed in terms of their quantities, accessibility, and formats.

- *Metadata*

  Metadata describe the collective characteristics of data. In short, metadata are *data about data*. Metadata details how, when, and where data has created or collected its documents scale, accuracy, resolution and other properties.

DATA AND DATABASES COMPONENTS

- ☑ A Digital Data Assessment and Review
- ☑ Master Data List
- ☑ Meta Data
- ☑ Critical Data Layers
  - Parcels
  - Address Points
  - Street Centerlines
  - Aerial Photography
- ☑ Departmental Specific Layers
- ☑ Enterprise Database Design (Example LGIM)
- ☑ Review of Database Design
- ☑ Data Creation Procedures
- ☑ Central Respository
- ☑ Custodianship (Data Stewards)
- ☑ Mobile Solutions
- ☑ Open Data/Open Government

FIGURE 3.4
Data and database components.

- *Critical data layers*

    In the context of geospatial technology, a *data layer* is the visual expression of accumulated data of a particular type. Elevation, city limits, or railway lines are all examples of data layers. Critical data layers refer to the data layers that are central to the GIS initiative.

    - *Parcels*

        A parcel is a legally defined area of land. A legal description of parcels of land for tax purposes.

    - *Address points*

        An address point is a location that is marked by its position relative to a physical structure. An address point is not necessarily the same as a street address. It is a data point that is assigned to a mapped location according to parameters that may or may not coincide with a street address.

    - *Street centerline*

        The street centerline is a linear data layer that correlates to the center of the roadway.

    - *Aerial photography*

        Aerial photography describes a bird's-eye-view style of photographic data that are gathered from a plane-, drone-, or helicopter-mounted camera. Because aerial photography produces an actual image of the mapped terrain, it improves the comprehensibility of practical details.

    - *Department-specific layers*

        Department-specific layers are mapped representations of data that correlate to the goals and objectives of a single department. For example, the position of every firehouse in a municipality would be *departmentally specific* to the city's emergency response services.

        Often, there are hundreds of digital data layers within an organization. Each layer can often be very specific to a department.

- *Enterprise database design*

    Enterprise database design refers to the way that an organization crafts a data repository in order to meet objectives and further the goals of the organization. Enterprise database design usually includes focusing on the data, the use of data models (Esri's Local Government Information Model [LGIM]), and integration strategies. The design specifies how an organization will collect, share, and act upon the various data to produce the desired information products.

- *Review of database design*

  This is the task of reviewing an organizational current database design and considering a migration to standardized models. The review should examine the breadth and efficacy of current technology.

- *Data creation procedures*

  Data creation procedures are the standardizing guidelines by which an organization's data are collected, cataloged, and turned into information products. This is an important set of procedures, as it protects against redundancy and needless work, both of which reduce overall cost-effectiveness.

- *Central repository*

  A central repository is an organization's aggregated collection of new and existing GIS data, gathered from all information resources. Pooling data in this manner allows for ease of maintenance, monitoring, and collection of metadata. A central repository of GIS data is a characteristic of an enterprise solution.

- *Custodianship (data stewards)*

  Data stewards are responsible for the administration and upkeep of specific digital data layers. They are custodians in that they monitor the accuracy as well as the security of departmental data.

- *Mobile solutions*

  Mobile solutions refer to GIS applications that are made available to users via a mobile device. In this day and age, mobile solutions are generally geared toward tablet and smartphone users.

- *Open data/open government*

  Open data and open government describe an increasingly prevalent policy that allows citizens, stakeholders, and nonstakeholders access to an organization's GIS-based data and data layers. Taxpaying citizens can see the results of a GIS initiative. Thus, a more transparent, open government is the end goal of this policy.

Figure 3.5 illustrates the end product after a self-assessment of the existing data and database conditions of the organization.

3.5.3 Category Three: GIS Procedures, Workflow, and Integration

Category three: GIS procedures, workflow, and integration is described by the following tasks. Figure 3.6 illustrates all of the key procedures, workflow, and integration components of an enterprise, sustainable, and enduring GIS solution in local government.

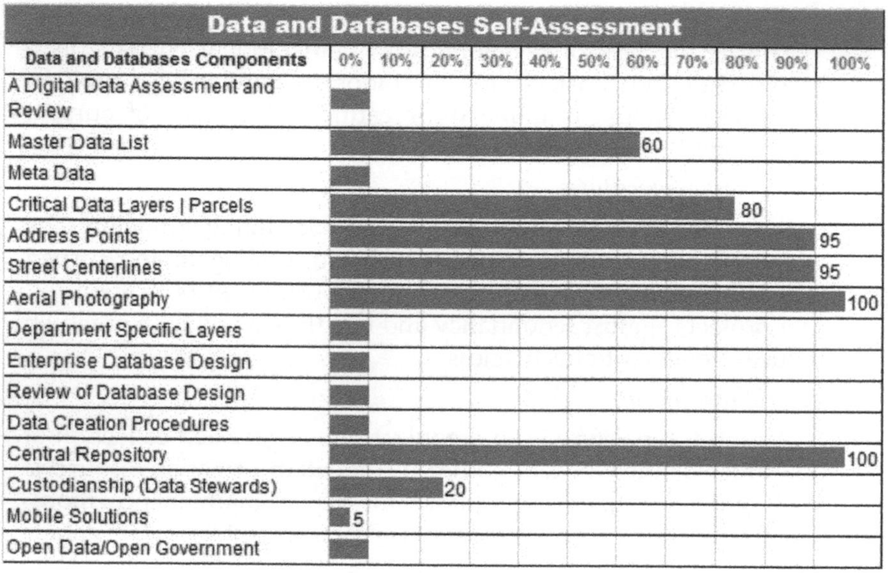

Data and Databases Self-Assessment												
Data and Databases Components	0%	10%	20%	30%	40%	50%	60%	70%	80%	90%	100%
A Digital Data Assessment and Review	■										
Master Data List							60				
Meta Data	■										
Critical Data Layers | Parcels									80		
Address Points										95	
Street Centerlines										95	
Aerial Photography											100
Department Specific Layers	■										
Enterprise Database Design	■										
Review of Database Design	■										
Data Creation Procedures	■										
Central Repository											100
Custodianship (Data Stewards)			20								
Mobile Solutions	5										
Open Data/Open Government	■										

FIGURE 3.5
Data and database self-assessment.

PROCEDURES, WORKFLOW AND INTEGRATION COMPONENTS

- ☑ Enterprise Integration
- ☑ Opportunities and Gaps
- ☑ Departmental Access to Critical Data Layers
- ☑ GIS Standard Operating Procedures (SOP)
- ☑ Data Maintenance Procedures
- ☑ GIS Application Acquisition/Development Procedures
- ☑ Meta Data Standards Defined
- ☑ Data Duplication Between Systems
- ☑ Level of Integration and Interoperability
- ☑ Work Order Solutions
- ☑ ERP Solutions (Permitting)
- ☑ Public Safety
- ☑ Enterprise Rather Than Departmental Silos
- ☑ GIS Technical Support (Ticketing/Help Desk)
- ☑ Departmental Use of GIS

FIGURE 3.6
Procedures, workflow, and integration components.

What do we mean by GIS procedures, workflow, and integration? *Procedures* denote a standardized method for accomplishing a designated task. *Workflow* describes the sequence of motions that a work item must undergo, from ideation to inception to completion. Workflows can encompass processes that span all levels of a department, from field collection to executive approval.

Integration refers to the process of taking disparate systems, and seamlessly integrating them into a single unit.

- *Enterprise integration*

  Enterprise integration describes the process whereby smaller disparate systems are integrated into the corporate initiative. In a geospatial context, enterprise integration encompasses not only the ways that information move from departmental systems to the central data repository, but also the ways that the new and larger system will alter stakeholder relationships and responsibilities.

- *Opportunities and gaps*

  Gaps in the enterprise and integrated GIS solution need to be identified and documented. It could include public safety data, permitting data, work order data, or crowdsourcing information. Opportunities are those databases that can effectively be incorporated into the enterprise GIS initiative.

- *Departmental access to critical data layers*

  Critical departmental data layers are crucial to the GIS enterprise. Departmental access refers to the ease with which various organizational departments may access these layers. Departmental accessibility is a critical component for success.

- *GIS standard operating procedures*

  Standard operating procedures (SOPs) are an organization's formally ratified blueprint for actions to be taken in pursuit of a desired objective. They are step by step, formulaic, and repeatable. In the geospatial context, *SOPs* prevent redundancy in data compilation and unnecessary effort. Adoption of SOPs also decreases organizational liability.

  - *Data maintenance procedures*

    Data maintenance procedures are a subset of SOPs that designate how to monitor and keep current the massive amounts of data that are collected in an enterprise GIS.

- *GIS application acquisition/development procedures*

  A GIS application simply refers to the deployment of GIS technologies to generate an information product. GIS application acquisition/ development procedures are a subset of SOPs detailing the ways in which GIS technologies are to be manipulated in order to meet user needs. How does a local government manage software acquisition and or custom software development? Do these procedures exist? Are they well documented? Does the organization understand the pros and cons of software development?

- *Metadata standards defined*

  It is critically important to define metadata standards. Metadata raise political as well as practical issues for enterprise GIS. Clear lines of accountability and quality control for the gathering, storage, and application of metadata should be ratified by an organization.

- *Data duplication between systems*

  Data duplication is the actual duplication of data layers. The most common GIS data layers that are duplicated in local government are street centerlines, address points, parcels, and, to a lesser extent, boundary layers. Some data layers existed in the databases of three separate departments; however, with the implementation of enterprise GIS, those duplicate data layers would be reduced to a single data layer within the central GIS database.

- *Level of integration and interoperability*

  The level of integration and interoperability measures how easily technological systems can share, interpret, and present data. An effective and enterprise GIS should integrate all databases and offer extensive interoperability. Interoperability means the ability of the GIS to work with other systems within and across organizational boundaries. This includes local, state, and federal data sources. The following is a list of key local government enterprise software solutions that require GIS integration:

  - *Work order solutions*

    As the name indicates, work order solutions manage, process, and maintain data about work orders and work that is performed. Work order solutions embrace asset management and GIS-centric solutions.

  - *Enterprise resource planning solutions (permitting)*

    Enterprise resource planning (ERP) solutions are integrative software applications that automate various functions that are related to planning, permitting, finance, and administration.

  - *Public safety solutions*

    Public safety solutions is the software application that is used in computer-aided dispatch, records management system, and other database and analysis tools.

- *Enterprise rather than departmental silos*

  Departmental silos are databases that are exclusively maintained by a single department. They are full of information and, like actual silos, vertically orientated but spread out over the terrain of an organization. For example, in a situation with departmental silos, the department of public safety may be the only department that keeps data on crime statistics. In an *enterprise* situation, however, all organizational departments have access to crime statistics via the central database that integrates all departmental data into a single master database.

- *GIS technical support (ticketing/help desk)*

  Like users of any IT, GIS users often need help or encounter problems while navigating GIS technologies. The team responsible for an organization's GIS technical support will walk users through issues and provides readily available troubleshooting information.

- *Departmental use of GIS*

  This is the actual utilization of GIS within all departments of local government. In the context of geospatial technology, departmental use implies a decentralized implementation of GIS technologies. This component should examine how effectively the departments are deploying the technology for different ends.

Figure 3.7 illustrates the end product after a self-assessment of the existing procedures, workflow, and integration conditions of the organization.

3.5.4 Category Four: GIS Software

GIS software refers to the network of programs and applications housed on mainframes, servers, and the cloud that are deployed to analyze, present, and draw conclusions from geospatial data. The end user interfaces with GIS technology via this software.

Figure 3.8 illustrates all of the key GIS software components of an enterprise, sustainable, and enduring GIS solution in local government.

- *Enterprise license agreement*

  A license agreement is a legal agreement that is entered into by the organization and a GIS software vendor that stipulates the limitations, liabilities, and appropriate applications of the vendor's technology. An enterprise license agreement (ELA) permits the deployment of a software program that is both organization-wide and ceiling-less in terms of user, data, or hardware restrictions. The objective here is to measure how available and pervasive GIS software is throughout the organization and create an optimum and cost-effective software deployment strategy.

- *Level of GIS commercial off-the-shelf versus custom code*

  GIS COTS is a GIS software that is a commercial off-the-shelf (COTS) software. It would seem that Esri is the *de-facto* local government standard and offers a comprehensive tool set for towns, cities, and counties. The objective is to evaluate how effective a

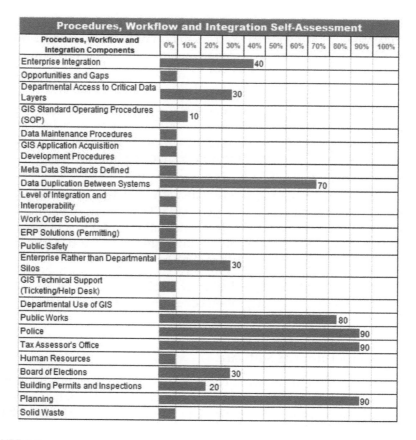

Procedures, Workflow and Integration Self-Assessment												
Procedures, Workflow and Integration Components	0%	10%	20%	30%	40%	50%	60%	70%	80%	90%	100%	
Enterprise Integration					40							
Opportunities and Gaps												
Departmental Access to Critical Data Layers				30								
GIS Standard Operating Procedures (SOP)		10										
Data Maintenance Procedures												
GIS Application Acquisition Development Procedures												
Meta Data Standards Defined												
Data Duplication Between Systems								70				
Level of Integration and Interoperability												
Work Order Solutions												
ERP Solutions (Permitting)												
Public Safety												
Enterprise Rather than Departmental Silos				30								
GIS Technical Support (Ticketing/Help Desk)												
Departmental Use of GIS												
Public Works									80			
Police										90		
Tax Assessor's Office										90		
Human Resources												
Board of Elections				30								
Building Permits and Inspections			20									
Planning										90		
Solid Waste												

FIGURE 3.7
Procedures, workflow, and integration self-assessment.

GIS SOFTWARE COMPONENTS

- ☑ Enterprise License Agreement (ELA)
- ☑ Level of GIS COTS Versus Custom Code
- ☑ Widget Development
- ☑ Access to Software
- ☑ Intranet Solution
- ☑ Public Access Portal
- ☑ Online Initiative
- ☑ Simple and Effective GIS Communication (Story Maps)
- ☑ Crowdsourcing Applications
- ☑ City Council GIS
- ☑ Modeling Extensions
- ☑ Mobile Software
- ☑ Global Positioning System (GPS) Technology

FIGURE 3.8
GIS software components.

local government organization is in using COTS versus developing custom GIS code. Open source code is, as the name implies, a code that can be used to create applications for and by local government. This would be a custom code strategy. However, GIS consultants use open source code to develop solutions that are essentially COTS. It is important to know how the extent of COTS versus custom code.

- Widget development

    A widget is a term for a small software program that augments the functionality of a larger software program. GIS widgets provide a way to customize applications in accordance with the specific needs and circumstances of an organization.

- *Access to software*

    In a geospatial context, access to software describes who can interact with what software, and to what extent. The objective is to evaluate how much access there is to GIS software within the organization.

- *Intranet solution*

    An Intranet is a web-based GIS solution that is accessible only to an organization's employees. A GIS Intranet solution is housed on a local government private network, accessible only to an organization's staff. Deploying Intranet GIS technologies also makes managing the software easier.

- *Public access portal*

    A public access portal is a Website where members of the public are able to interact with GIS information. Public web portal inform and enable citizens.

- *Online initiative*

    An online initiative is a program that is established to plan, design, and deploy cloud-based GIS solutions. Esri's ArcGIS Online is a leading solution for local government and offers cloud-based alternatives.

- *Simple and effective GIS communication (story maps)*

    Simple and effective Web-based maps that tell a story of an event, history, or occasion within local government can have a powerful effect on the community.

- *Crowdsourcing applications*

    Crowdsourcing is a twenty-first-century process of obtaining resources through an online community. Crowdsourcing applications are software programs that facilitate the interaction between organizations and the online community.

- *City council GIS*

  City council GIS refers to the use of GIS technology by elected officials to view geographic information about the various issues that are related to the organization.

- *Modeling extensions*

  A modeling extension is similar to a widget, in that it is a specialized software that helps organizations with specific business and operational needs. A modeling extension is larger in scale than a widget and enhances the overall representative capabilities of a program. An example would be a routing and scheduling algorithm modeling extension.

- *Mobile software*

  Mobile software refers to GIS applications that are designed for mobile use on a tablet or a smartphone. The mobility of GIS is a critical component of any successful enterprise GIS.

- *Global Positioning System technology*

  Global Positioning System (GPS) technology is a navigational system that is enabled by a network of satellites orbiting the earth. The satellites are constantly broadcasting their positions in the sky so that a GPS receiver on earth can pick up these signals and self-triangulate according to the information that is received. Though people often get the GIS and GPS confused, GPS is a single, though important, tool on the belt of GIS technology. GPS can be used for gathering and monitoring geospatial data.

Figure 3.9 illustrates the end product after a self-assessment of the existing GIS software conditions of the organization.

3.5.5 Category Five: GIS Training, Education, and Knowledge Transfer

GIS training is the action of *teaching a particular* skill or a new type of behavior. Training tends to be more formal and often includes computer technology. GIS education is the *enlightened experience* that follows systematic instruction and usually occurs in an academic setting. Education is less formal than GIS training and does not include anything but the student's presence. GIS knowledge transfer is the art of *transferring knowledge* from one part of the organization to another. This is usually accomplished in a very relaxed atmosphere.

Figure 3.10 illustrates all of the key GIS training, education, and knowledge transfer components of an enterprise, sustainable, and enduring GIS solution in local government.

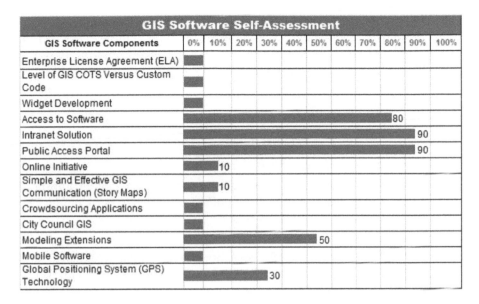

GIS Software Self-Assessment											
GIS Software Components	0%	10%	20%	30%	40%	50%	60%	70%	80%	90%	100%
Enterprise License Agreement (ELA)											
Level of GIS COTS Versus Custom Code											
Widget Development											
Access to Software									80		
Intranet Solution										90	
Public Access Portal										90	
Online Initiative	10										
Simple and Effective GIS Communication (Story Maps)	10										
Crowdsourcing Applications											
City Council GIS											
Modeling Extensions					50						
Mobile Software											
Global Positioning System (GPS) Technology			30								

FIGURE 3.9
GIS software self-assessment.

TRAINING, EDUCATION AND KNOWLEDGE TRANSFER COMPONENTS

- ☑ Formal On-Going GIS Training Plan
- ☑ Multi-Tiered GIS Software Training
- ☑ Mobile Software Training
- ☑ Departmental Specific Education
- ☑ ROI Workshops
- ☑ Knowledge Transfer
- ☑ Conferences
- ☑ Online Seminars and Workshops
- ☑ Brown Bag Lunches
- ☑ Succession Planning

FIGURE 3.10
Training, education, and knowledge transfer components.

The following list details the GIS training, education, and knowledge transfer components of an enterprise GIS:

- *Formal ongoing GIS training plan*

  A formal ongoing GIS training plan is a ratified outline of steps, schedules, and costs for continuing to train an organization's employees. It is important to have an ongoing training plan, considering that GIS is a rapidly evolving technology, and organizational needs are ever changing.

- *Multitiered GIS software training*

  Multitiered GIS software training refers to a standardized process for training employees in the use of GIS technology. Multitiered training is defined by four distinct types of GIS users, as described in Chapter 7.

- *Mobile software training*

  Mobile software training is the process of teaching users how to engage with GIS technology on their mobile device.

- *Departmental-specific education*

  Departmental-specific education provides specialized training procedures according to a department's specific needs. See Chapter 7 for departmental-specific training.

- *RoI workshops*

  RoI workshops are specific workshops that are related to the value and RoI that GIS offers the enterprise. Each department is an important component in the success of an enterprise GIS.

- *Knowledge transfer*

  Knowledge transfer refers to the process of communicating the GIS know-how and knowledge among different entities in an organization. Knowledge transfer is the art of transferring knowledge from one part of the organization to another.

- *Conferences*

  Conferences are gatherings of the GIS community that provide a smorgasbord of opportunities for furthering employee GIS education. Talks, lectures, lessons, and socialization with other industry professionals are ways to advance an understanding of geospatial technologies and keep abreast of new developments.

- *Online seminars and workshops*

  Online seminars and workshops are online programs that are implemented by a variety of organizations that further GIS education among employees.

- *Brown bag lunches*

  Brown bag lunches are as informal as they sound. This term refers to a free-and-easy meeting, generally held over a meal, where employees can discuss concerns with GIS in a social setting.

- *GIS succession planning*

  Succession planning refers to an organization's strategy for filling essential but vacant positions with experienced employees.

Figure 3.11 illustrates the end product after a self-assessment of the existing GIS training, education, and knowledge transfer conditions of the organization.

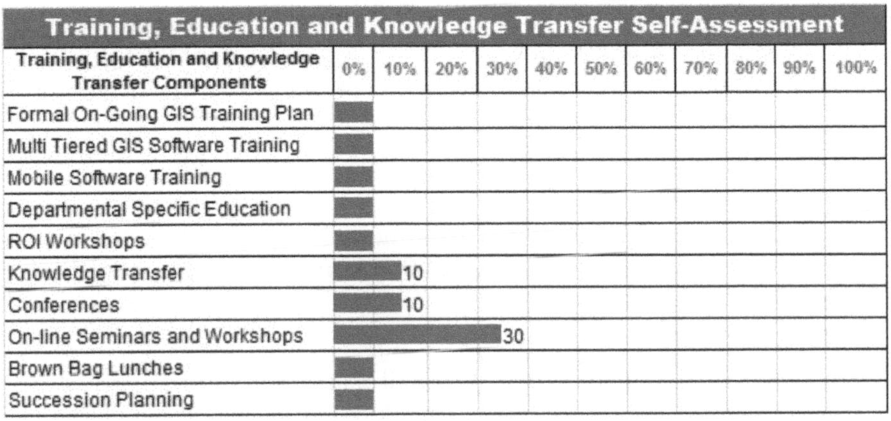

Training, Education and Knowledge Transfer Self-Assessment											
Training, Education and Knowledge Transfer Components	0%	10%	20%	30%	40%	50%	60%	70%	80%	90%	100%
Formal On-Going GIS Training Plan											
Multi Tiered GIS Software Training											
Mobile Software Training											
Departmental Specific Education											
ROI Workshops											
Knowledge Transfer	10										
Conferences	10										
On-line Seminars and Workshops		30									
Brown Bag Lunches											
Succession Planning											

FIGURE 3.11
Training, education, and knowledge transfer self-assessment.

3.5.6 Category Six: GIS Infrastructure

Infrastructure refers to the network of structures, both physical and systemic, that support an organization's GIS activity. Figure 3.12 illustrates all of the key GIS infrastructure components of an enterprise, sustainable, and enduring GIS solution in local government.

The following details the GIS infrastructure components of an enterprise GIS:

- *Strategic technology plan*

    A strategic technology plan describes an organization's current and future relationship with technology and outlines how this technology will further the goals of the organization.

GIS INFRASTRUCTURE COMPONENTS

- ☑ Strategic Technology Plan
- ☑ GIS Architectural Design
- ☑ IT Infrastructure
- ☑ IT Replacement Plan
- ☑ GIS Training for IT Professionals
- ☑ 24/7 Availability
- ☑ Enterprise Back-Up
- ☑ Data Storage
- ☑ IT, Hardware and Mobile Standards
- ☑ GIS Mobile Action Plan
- ☑ GIS Staging and Development Zone

FIGURE 3.12
GIS infrastructure components.

- *GIS architectural design*

    GIS architectural design is the plan that addresses GIS software technology, capacity performance, and IT infrastructure including hardware, network communications, software architecture, enterprise security, backup, platform performance, and data administration.

- *IT infrastructure*

    IT infrastructure refers to a dynamic web of processes, networks, hardware, and software resources that support the activities of an integrated IT department.

- *IT replacement plan*

    An IT replacement plan is a formal plan for updating hardware and software resources in the future. Budgetary concerns, goals, and long-term objectives are taken into account.

- *GIS training for IT professionals*

    In order for IT professionals to assist an organization with many GIS activities including but not limited to *crowdsourcing* or *tech support*, they need a proficiency in GIS technologies.

- *24/7 availability*

    The term 24/7 availability refers to the availability of IT infrastructure and GIS technology at all hours of the day, every day of the week.

- *Enterprise backup*

    Enterprise backups are a protective measure that preserves an organization's centralized data through off-site cloud-based daily backup procedures.

- *Data storage*

    Data storage refers to the digital information storage locally and on the enterprise network and in the cloud.

- *IT, hardware, and mobile standards*

    IT, hardware, and mobile standards refer to the formalized set of guidelines and requirements that are required by the organization to support an enterprise GIS.

- *GIS mobile action plan*

    A mobile action plan is an outline of the tactics that an organization will deploy in order to increase GIS accessibility on tablets and smartphones.

- *GIS staging and development zone*

    A development zone is a site where newly developed GIS applications are tested and tweaked. A staging zone is a site where GIS applications are given full-trial runs.

GIS Infrastructure Self-Assessment											
GIS Infrastructure Components	0%	10%	20%	30%	40%	50%	60%	70%	80%	90%	100%
Strategic Technology Plan											
GIS Architectural Design											
IT Infrastructure			25								
IT Replacement Plan										90	
GIS Training for IT Professionals				35							
24/7 Availability											100
Enterprise Back-Up											100
Data Storage											100
IT, Hardware and Mobile Standards											
GIS Mobile Action Plan											
GIS Staging and Development Zone											

FIGURE 3.13
GIS infrastructure self-assessment.

Figure 3.13 illustrates the end product after a self-assessment of the existing IT infrastructure conditions of the organization.

3.6 The Formula for Success

Sections 3.5.1 through 3.5.6 illustrate my idea of a simple and understandable checklist of all the critical components and tasks that are required for a successful and enterprise GIS. If we are going to assign a value of completeness for each of the categories that are listed in Sections 3.5.1 through 3.5.6, it is important to have the utmost clarity on the meaning of that task. The information should help you understand what each component means. This will help you grade yourself on your organization's achievement level. The objective is to literally create a *heartbeat* of your organization.

Figure 3.14 presents an entire GIS strategy on the back of an envelope and represents the outcome of completing the self-evaluation by using an online questionnaire that I developed in 2016.

This online questionnaire offers the following to local government professionals:

- Statistical summary of findings
- GIS strengths
- GIS weaknesses
- GIS threats

- GIS opportunities
- A possible GIS road map for the future
- Estimated budget in dollars and person hours

The following represent the documented findings and recommendations after using the formula for success. We were able to create a vision, goals, and objectives for these organizations; identify weaknesses; and essentially create a road map for improvement.

3.6.1 Governance Recommendations

Vision: Implement an optimum GIS governance model that centralizes technology and decentralizes users.

Goal: Provide management with understandable strategies for the effective utilization of GIS technology. In addition to improving overall governance, this strategy should include clear lines of responsibility and facilitate stakeholder's decision making.

Task 1: Adopt the GIS SIP

Adopt and fund the GIS SIP. Remember, this is the GIS SIP that is developed according to the steps that are laid out in Chapter 2.

Task 2: Annual updates to the GIS SIP

Update the GIS SIP on an annual basis. Modified versions of the *online questionnaires* and *departmental interviews* should be deployed, and the organization should benchmark its successes using the methodology that is laid out in the GIS SIP.

Task 3: GIS vision, goals, and objectives

Adopt and formally ratify the visions, goals, and objectives that are laid out in the GIS SIP to ensure organization-wide commitment and understanding.

Task 4: A formalized governance model

Adopt and ratify an enterprise governance model that supports a sustainable GIS.

Task 5: Job classifications

Modify the job descriptions and titles for staff who are involved with the implementation of GIS technology. Job titles are important designators for staff, and deploying the correct label can have a positive effect on personal accountability and technological engagement.

Task 6: Enterprise GIS project management

Emphasize a governance model and staffing structure that centers on GIS technology management and support.

Task 7: Coordinated GIS enterprise

Emphasize a governance model and staffing structure that centers on the coordination of the GIS enterprise.

Task 8: GIS steering committee

Continue to support its GIS steering committee as it manages and maintains the GIS technologies. The steering committee will also serve in the role of a GIS technical committee.

Task 9: GIS sponsor team

Have a team of sponsors for the initiative, headed by the GIS planning department director, and the senior management and the steering committee.

Task 10: GIS technical committee

Establish the responsibilities of the technical committee among members of the steering committee.

Task 11: GIS functional groups

Consider GIS functional teams in the third year of the initiative. These functional teams may be in the public safety, land administration, and utility departments.

Task 12: GIS user group

Formalize, promote, and organize a GIS user group that will meet on a monthly or quarterly basis to discuss user-specific issues.

Task 13: Regionalization of GIS

Work closely with external entities, regional consortiums, and adjacent organizations in order to promote coordination, save resources, and foster cooperative attitudes.

Task 14: GIS policy and mandates

Formalize a GIS data-sharing and dissemination policy that includes disclaimers on data quality.

Task 15: User sensitivity

Conduct an annual survey via online questionnaire in regard to cross-departmental GIS user needs. This survey should be planned through the leadership of the GIS initiative.

Task 16: GIS collaboration

Coordinate and administer meetings, workshops, seminars, and project-specific discussions to maximize interdepartmental collaboration and opportunities for data integration.

Task 17: Measure quality of service

Distribute an online questionnaire inquiring about the quality of the services as outlined in the departmental SLA.

Task 18: GIS authority and clear lines of responsibility

Support ratified governance model with clear and concise lines of responsibility, accountability, authority, and custodianship of the digital data layers.

Task 19: A GIS budget or funding model

Fund and sustain the future enterprise GIS using annual operating funds.

Task 20: Chargeback model

Replace the existing chargeback model with unified government operating funds.

Task 21: Grants and funding initiatives

Investigate and identify all research and funding opportunities at the local, state, and federal levels.

Task 22: An annual detailed GIS work plan and SLA

Develop a detailed work plan based upon the SLA that is endorsed by the GIS steering committee.

Task 23: GIS coordination tasks

Coordinate and manage all interdepartmental and regional GIS projects.

Task 24: KPIs

Monitor and grade the performance of the enterprise GIS initiative by benchmarking its achievement of the annual KPIs that are indicated in the GIS SIP. The lead GIS person will be ultimately responsible for presenting the success of the GIS initiative, as well as detailing challenges, barriers, and pitfalls.

Task 25: GIS blog or newsletter

Develop a quarterly GIS blog and newsletter, and link it to other social media platforms such as Facebook, Twitter, and LinkedIn.

Task 26: A GIS culture of collaboration

Create a culture of collaboration throughout the unified government organization and all external entities. Accomplish this by improving communication and branding of the GIS enterprise initiative while also promoting activities that foster collaboration between departments, or divisions within departments.

Task 27: Alignment with organization's vision, goals, and objectives

Align all the GIS activities that are stated in the three-year GIS SIP, SLA, and action plan with the overall mission and vision of the organization.

3.6.2 Data and Databases

Vision: Design, build, update, collect, and maintain reliable and sustainable GIS digital and data layers.

Goal: Use Esri's LGIM as the standardized data model for future growth. A modified LGIM should be used to build and maintain accurate, consistent, and reliable geographic data.

Task 1: Digital data assessment

Perform a comprehensive assessment of the quality, quantity, and completion of *all* digital data layers. The organization needs to supplement their existing digital data repository with new local, state, and federal data.

Task 2: Master data list

Create, update, and maintain a master digital data list.

Task 3: Metadata

Establish and enforce SOPs for metadata-related activities.

Task 4: Parcels (critical data layer)

Improve the accuracy and reliability of the digital parcel layer, based on the digital data assessment.

Task 5: Address points (critical data layer)

Improve the accuracy and reliability of the digital address point layer, based on the digital data assessment.

Task 6: Street centerlines (critical data layer)

Improve the accuracy and reliability of the digital street centerline layer, based on the digital data assessment.

Task 7: Aerial photography (critical data layer)

Improve the accuracy and reliability of the digital aerial photography layer, based on the digital data assessment.

Task 8: Enterprise database design (LGIM)

Migrate their existing enterprise digital repository to Esri's LGIM and thus allow opportunities for customization.

Task 9: Review of database design

Review the newly proposed LGIM database design. All organizational stakeholders should complete this task.

Task 10: Data creation procedures

Develop uniform standards for the creation of all new digital GIS data. Essentially, this implies a new formal SOP and may include multiple methods for new data creation.

Task 11: Central repository

Create a central repository for all enterprise digital GIS data layers. Security and permission settings will allow the departmental ownership and custodianship of each and every digital data layer. The organization should anticipate using Esri's ArcSDE environment to house all GIS data. Exceptions to this rule will be documented, detailed, and authorized by the lead GIS person.

Task 12: Custodianship

Clearly define departmental data custodianship roles within the enterprise governance model. Include an agreement between all departments regarding the lines of responsibility for every digital data layer.

Task 13: Mobile solutions for viewing and maintaining data

Plan, design, and deploy Esri's ArcGIS Online as the organization's mobile software solution. Supplement this technology with continuing efforts to identify advanced, convenient, and easy-to-use mobile GIS and GPS field tools to collect, update, and maintain the organization's GIS data repository. A uniform approach to using ArcGIS would benefit the organization.

Task 14: Open data government

Develop an open data initiative based on the best business practices from other organizations. This initiative is focused on the responsible sharing of information with citizens and other organizations.

3.6.3 Procedures and Workflow

Vision: Promote the interoperability of GIS with existing business systems.

Goal: Integrate GIS functionality with existing database systems, business processes, and workflow.

Task 1: Enterprise integration

Integrate GIS with the organization's existing business systems.

Task 2: Opportunities and gaps

Identify existing gaps and opportunities for GIS integration, and continue to do so on a regular basis.

Task 3: Departmental access to critical data layers

Monitor stakeholder access to all critical digital data layers. Accomplish this task through the monitoring of software use and communication with all stakeholders. These steps will allow improved feedback on the completion and accuracy of digital data.

Task 4: Develop enterprise SOPs

Establish a set of standards and procedures for managing, maintaining, and updating geospatial data. These standards could include, but are not limited to, the following:

- Office-to-field/field-to-office procedures
- GPS quality standards
- GIS versioning
- Computer aided design (CAD) standards
- Digital submission standards
- Cartographic standards
- Metadata standards
- Standard naming conventions
- GIS business integration
- Parcel fabric maintenance

Task 5: Data maintenance procedures

Develop all procedures and protocols for the maintenance of digital data by each departmental custodian. This will include multiple techniques for data maintenance.

Task 6: GIS application development

Develop policies on and standards for any new application development projects including custom browsers, widgets, and any business productivity-enhancing tools that are developed in-house.

Task 7: Define metadata standards

Develop and define what is expected from the metadata. This is the responsibility of the GIS steering/technical committee.

Task 8: Data duplication between systems

Avoid data duplication between the various systems. The organization's lead GIS person should develop procedures and protocols to achieve this.

Task 9: Work order solutions

Evaluate and determine the efficacy of GIS integration with the organization's work order solution.

Task 10: ERP solutions

Evaluate and determine the efficacy of GIS integration with the organization's ERP solution.

Task 11: Permitting

Evaluate and determine the efficacy of GIS integration with the organization's permitting solution.

Task 12: Public safety

Evaluate and determine the efficacy of GIS integration with the organization's public safety solution.

Task 13: Enterprise rather than departmental silos

Develop a true enterprise solution rather than constraining its efforts in departmental silos. The organization's lead GIS person should constantly evaluate the enterprise versus silo balance within the organization.

Task 14: GIS technical support

Manage and maintain technical support to all departments. Include a ticketing system to monitor the quality of GIS technical support.

Task 15: Departmental GIS

Monitor the effective use of GIS by each department. The organization's lead GIS person will be responsible for presentations on the GIS-related increases in efficiency. Anticipate each department becoming less reliant on the central GIS staff for map production, analysis, and decision support.

3.6.4 GIS Software

Vision: Make GIS software accessible throughout the organization and to the public.

Goal: Deploy a full suite of Esri GIS software solutions (desktop, Intranet, Internet, and mobile) across the enterprise.

Task 1: ELA

Evaluate ELA versus a single seat license. Maximize the efficacy of Esri's software by deploying the right tools to the right people.

Task 2: Level of GIS code versus custom code

Monitor the use of custom applications versus the Esri suite of software products. Anticipate that the organization will utilize Esri's suite of solutions for the enterprise, which will ultimately limit the development or purchase of custom software solutions.

Task 3: Access to software

Make GIS solutions available to all stakeholders and citizens within the enterprise.

Task 4: Intranet solution

Utilize Esri's ArcGIS online (AGOL) Map Viewer as the Intranet portal of choice. This includes the planning, design, and customization of Esri's HTML5 browser template.

Task 5: Public access portal

The organization should continue to use the existing solution as their access portal. Furthermore, the organization should take advantage of Esri's AGOL Internet Map Viewer to consider alternative solutions in the future. Specific departmental opportunities for the development of Esri-based public access portals will present themselves.

Task 6: Online initiative

Plan, design, and deploy AGOL. Include the setup, configuration, and effective use of the tools and applications that are made available in the Esri licensing agreement.

Task 7: Simple and effective GIS communication (story maps)

Develop a sequence of story maps according to the following five-step process:

1. Storyboard
2. Gather data
3. Design
4. Build and refine
5. Publish and maintain

Task 8: Crowdsourcing applications

Consider a crowdsourcing application. This should be the responsibility of the steering/technical committee.

Task 9: Organizational GIS

Enable the organization with GIS technology and utilize real-time GIS software in meetings.

Task 10: Modeling extensions

Take advantage of Esri's modeling extensions for the desktop.

Task 11: Mobile software

Plan, design, and deploy Esri's AGOL as the organization's mobile software solution. Use this solution with a continued effort for new, advanced, and convenient mobile GIS and GPS field tools to collect, update, and maintain the organization's GIS data repository.

Task 12: GPS technology

Continue to evaluate when and where GPS technology should be used.

3.6.5 GIS Training, Education, and Knowledge Transfer

Vision: Train, educate, and promote knowledge transfer among all staff members.

Goal: Improve the GIS knowledge base within organizational departments. Develop a training plan that promotes effective knowledge transfer. Encourage the effective utilization of GIS technology.

Task 1: Formal ongoing training plan

Implement a formal, sustainable GIS training plan.

Task 2: Multitiered GIS software training

Provide GIS training to staff on a regular basis. Utilize Esri's online education and training services and provide formal classroom training for identified departmental staff. Include desktop, Intranet, Internet, mobile, GPS, ArcGIS Online, collector application, operations dashboard, automated vehicle location story maps, ArcGIS Pro, and extensions in the training procedures.

Task 3: Mobile software training

Develop a strategy for effective training on mobile field devices as part of the formal training plan.

Task 4: Departmentally specific education

Offer training and education workshops on a monthly basis during year one of the GIS initiative. Include the following:

1. *Introduction Series*
   a. Introduction to GIS
   b. GIS strategic planning
2. *Software Series*
   a. Esri software suite
   b. Local government business applications
   c. Mobility of GIS
   d. Public safety and emergency operations center
   e. Parks, lands, and natural resources
3. *Management Series*
   a. Business case for GIS (RoI)
   b. GIS architecture/hardware/software/communication
   c. GIS manager's workshop—best business practices for GIS
   d. GIS challenges
   e. The future of GIS
   f. GIS implementation plan/status report

Task 5: RoI workshops

Conduct an RoI workshop for all departments.

Task 6: Knowledge transfer

Establish the GIS user group network as a knowledge transfer opportunity.

Task 7: Conferences

Attend workshops and preconference seminars at the Esri International User Conference and regional Esri conferences, such as South East Region Users Group (SERUG).

Task 8: Online seminars and workshops

Utilize all available online training, education, and knowledge transfer workshops.

Task 9: Brown bag meetings

Offer seminars and workshops that are tailored to the specific departmental applications of GIS.

Task 10: Succession planning

Develop a strong GIS user base via training, hiring practices, and proactive succession planning.

3.6.6 Infrastructure

Vision: Continue to utilize the IT infrastructure to support an enterprise, scalable, and sustainable GIS.

Goal: Continually evaluate the organization's architecture initiative so that it will sustain enterprise growth and change.

Task 1: Develop a strategic technology plan

Participate in the multiyear IT strategic plan. The purpose of this plan is to document the existing and future IT conditions. IT is essential to understand this strategy as it relates to GIS and GIS implementation.

Task 2: Develop a GIS architectural system design

The GIS architectural assessment and system design should include the following:

1. *Executive summary*
   a. Central server and service-oriented infrastructure approach
2. *Methods, constraints, and acknowledgments*
   a. Purpose and methodology
   b. Assumptions and constraints
   c. Acknowledgments

3. *Architecture vision*
4. *Business architecture*
   a. Business requirements
   b. IT standards and policies
5. *Technology architecture*
   a. Applications
       i. Server
       ii. Desktop
       iii. Mobile
   b. Hardware
       i. Server
       ii. Desktop
       iii. Mobile
   c. Network communications
6. *System architecture design*
   a. Platform sizing
   b. Server software performance
       i. Process configuration
       ii. Cached map services
       iii. Memory configuration
   c. GIS data administration
       i. ArcSDE geodatabase
       ii. Local Government Information Model
       iii. GIS imagery data architecture
       iv. Storage architecture options
   d. Network communications
       i. Capacity and performance
       ii. Suitability analysis
   e. Platform performance
7. *GIS architecture recommendations*
   a. GIS server platform
   b. Desktop platform
   c. Mobile platform
   d. Network communications
   e. Best practices
   f. Standard operating procedures

Task 3: IT infrastructure

It is important to understand your organization's IT infrastructure. It includes hardware, software, network resources, and the services required for the existence, operation, and management of an enterprise IT and GIS environment.

Task 4: Infrastructure technology replacement plan

Understand the IT replacement strategy within your organizations. Integrate it into the multiyear GIS implementation strategy.

Task 5: GIS training for IT professionals

GIS training for the IT department is an important component of any GIS initiative. Training and educating IT staff will play an important role in your GIS deployment.

Task 6: Confirm 24/7 availability

The GIS needs to be available 24/7. Connect with the IT staff, and make sure that your organization has the IT infrastructure to support 24/7 availability.

Task 7: Enterprise GIS backup

Enterprise GIS backup is a prerequisite for any enterprise GIS deployment.

Task 8: GIS data storage

GIS has historically been a space hog. Confirm that your organization has the procedures and protocols and infrastructure for all the required data storage.

Task 9: IT, hardware, and mobile standard

Standards are a key component of any GIS initiative. Participate in your organization's IT standards.

Task 10: GIS mobile action plan

Develop a mobile action plan for GIS.

Task 11: GIS staging and development zone

Develop an area for the testing and staging of GIS solutions and services. Testing system components, database development, and maintenance activities are important.

Figure 3.14 illustrates your entire GIS road map on the back of an envelope.

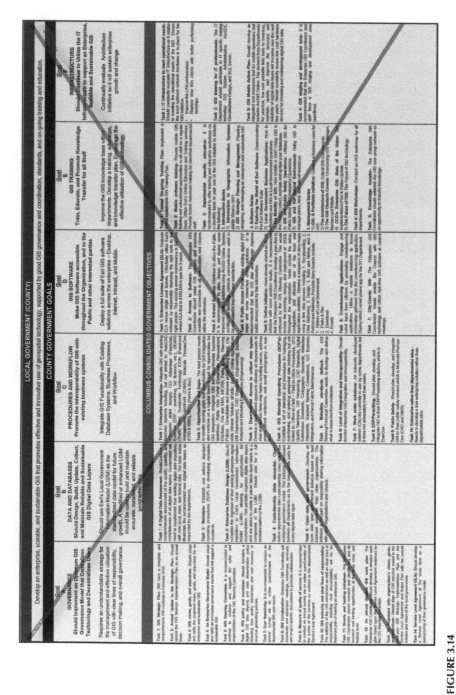

FIGURE 3.14

Your entire strategy on the back of an envelope.

4

Challenges, Barriers, and Pitfalls

The fault, dear Brutus, is not in our stars, but in ourselves.

William Shakespeare, *Julius Caesar*

4.1 Introduction

Understanding the reasons for failure will help us solve problems in the future. Therefore, it is important to shine a bright light on the reasons why local government often fails to create a truly enterprise, sustainable, and enduring GIS. I do hope that this chapter will at least stimulate further research on what it takes to deploy enterprise solutions. At the very least, we are starting to create actionable intelligence. I want to start and end this chapter with a provocative statement that should resonate with you all, namely, that an organization's failure to implement and maintain an enterprise, sustainable, and enduring GIS begins and ends with the failure of the GIS coordinator.

We must recognize that technology has changed significantly over the last few decades, and it is only now that enterprise solutions are technically within our grasp. Since the technology itself can handle anything that an organization throws at it, we can therefore presume that the challenges, the barriers, and the pitfalls are more about the members of an organization than anything else. This is known as the *human factor*.

In Daniel Gilbert's book *Stumbling on Happiness* (2006), he writes, *"A sophisticated machine could design and build any one of these things because designing and building require knowledge logic and patience, of which sophisticated machines have plenty. In fact, there's really only one achievement so remarkable that even the most sophisticated machine cannot pretend to have accomplished it, and that achievement is conscious experience."*

Figure 4.1 illustrates the level of complexity and components of an enterprise GIS.

It is this state of awareness of all of the moving parts of a GIS and the decision-making complexities, whether internal or external, that can get us into GIS management trouble. Essentially, it is this human side, or the outcomes of our conscious experience of GIS, that creates the challenges,

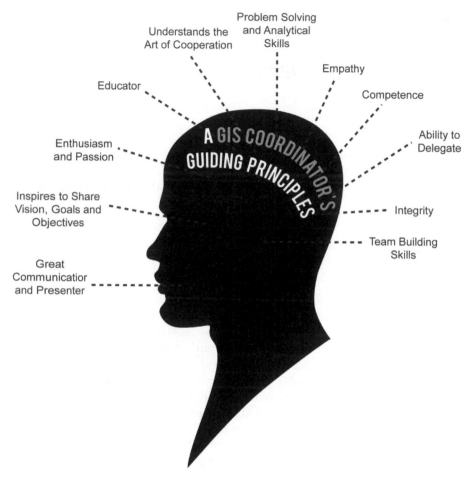

FIGURE 4.1
Conscious experience.

barriers, and pitfalls in our path. So, keep this in mind as we explore the GIS challenges, barriers, and pitfalls.

Firstly, we should take a look at a *pass/fail* assessment of the GIS implementation process. Secondly, we will compartmentalize the task of GIS implementation into five GIS planning components. Thirdly, we will determine what we exactly mean by challenges, barriers, and pitfalls. After all, what is a barrier, and how is it different from a challenge? Fourthly, we will take a holistic view of GIS implementation to identify where, along the way, the challenges, barriers, or pitfalls are likely to crop up. If a three-year time frame is used to implement an optimum GIS, when and where should we expect problems?

4.2 A Pass/Fail Approach

Let us start with a no-nonsense look at a pass–fail comparison of six local government organizations. Figures 4.2 through 4.4 contrast the successes of one organization against the failures of another. There are many moving parts in the enterprise GIS implementation process, and a problem in a single area can cause a ripple effect throughout the organization.

Figure 4.2 compares two city approaches to GIS implementation.

Figure 4.2 illustrates the checklist of components that need to be in place for success to be guaranteed. City B failed to embrace some of the most obvious components of a successful GIS. The organization tried to deploy GIS without a detailed plan. This is a show-stopping error in itself. City A checked off almost all of the major components that are required for success. As the old adage states, "Forewarned is forearmed;" thus, if an organization wants a passing grade, they need to address every one of the components that are detailed in the formula for success, not just the examples in the figure.

Figure 4.3 compares two county organizations' approach to GIS implementation.

How did County B doom its GIS implementation efforts? They made no concerted effort to educate staff or promote the technology. Elected officials were not educated or sold on the benefits of GIS. Readers should note that County Government B likely thinks that they have a perfect and enterprise solution. However, the failings of a GIS coordinator are transparent.

Figure 4.4 compares the different GIS implementation approaches of two small towns.

CITY A	CITY B
GIS Implementation - PASS	**GIS Implementation - FAIL**
Multi Year GIS Strategic Implementation Plan	No Strategic Implementation Plan or Direction
A Key Sponsor	No Key Sponsor
Optimum Staffing Plan	Staffed Incorrectly
Good Governance Model - GIS Housed in IT Department	**Poor** Governance Model - GIS Placed in Public Works Department
Enterprise Departmental Intranet deployed	No Enterprise-wide Solutions
Published Successes and Secured Awards	No Success Stories
Used State, Federal and Local data	Strict Data Accuracy Standards
Invested less than $1,000,000	Invested well over $1,000,000
Elected Officials support annual GIS budget	Elected Officials pulled funding for the GIS Initiative
Good GIS Coordinator	Poor GIS Manager

FIGURE 4.2
City A versus City B.

COUNTY A	COUNTY B
GIS Implementation - PASS	**GIS Implementation - FAIL**
Three-Year GIS Strategic Implementation Plan	Strategic Implementation Plan
A Key Sponsor	No Key Sponsor
Optimum Staffing Plan	Small Staff with limited power
Good Governance Model - GIS Housed in IT Department	Good Governance Model - GIS Placed in IT but too few staff
Enterprise Software Solutions deployed	Good Enterprise-wide Solutions
Some Publications	No Success Stories - Staff refused to change and adapt to GIS technology
Used all available digital data	Used all available digital data
Continued Investment	Reasonable Investment
Elected Officials support annual GIS budget	Elected Officials were not educated and sold on GIS
Fair Training, Education and Knowledge Transfer tools	Non-existent Training, Education and Knowledge Transfer tools
Good GIS Coordinator	Mediocore GIS Coordinator

FIGURE 4.3
County A versus County B.

TOWN A	TOWN B
GIS Implementation - PASS	**GIS Implementation - FAIL**
Three-Year GIS Strategic Implementation Plan	No Strategic Implementation Plan just a GIS Assessment without details
A Key Sponsor	No Key Sponsor
Optimum Staffing Plan by Outsourcing the GIS	One Key Staff
Good Governance Model - GIS Outsourced	Good Governance Model - Some outsourcing as needed
Enterprise Software Solutions deployed	Good Enterprise-wide Solutions
International Publications and Success Stories	Limited Success Stories
Used all available digital data	Used all available digital data
Annual Investment	Good Investment
Elected Officials support annual GIS budget	Elected Officials and Steering Committee were not educated on the benefits of GIS
Fair Training, Education and Knowledge Transfer tools	Limited Training, Education and Knowledge Transfer tools
Great GIS Project Manager/Coordinator - outsourced solutions	GIS Coordinator lacks sales and technical skills

FIGURE 4.4
Town A versus Town B.

Town A is a beautiful town in California that elected to outsource their GIS and implement an enterprise solution over four years. Like many smaller organizations, Town B lacks an understanding of what the GIS can really do for the organization and failed to solve issues of education and knowledge transfer.

The examples above are real organizations that I opted not to name, as there is never a good reason to embarrass any person or organization. Planning, senior support, staff, and good governance, all supported by enterprise tools and a willingness to promote positive stories, are critically important to the success of an enterprise GIS. Though there is a long list of ingredients that are required for GIS success, some are more important than others. Let us turn our attention to the five crucial components of a successful enterprise GIS.

4.3 Five GIS Strategic Planning Components

A client once asked me, what are the top five challenges in adopting and implementing a successful enterprise GIS? My answer needed plenty of wiggle room for details. I did not want to just say data, people, technology, and so forth. The response I came up with was fivefold. The five main categories are (1) strategic, (2) tactical, (3) technical, (4) logistical, and (5) political. Let us take a closer look at all of them.

Figure 4.5 introduces five components that are related to the challenges that are faced by many local government organizations.

4.3.1 Component One: Strategic GIS Components

Strategic components are the things that are methodically planned, deliberate, and calculated to further the *bigger picture* of the activity of a GIS

FIGURE 4.5
Five components.

initiative. These may include a GIS strategic plan, an annual GIS action plan, a service level agreement (SLA), or a memorandum of understanding (MU). Strategic components are essentially the navigation charts for your organization's voyage. Figure 4.6 is a list of the strategic components of an enterprise GIS. Strategic components should include answers to the following questions:

- What is the vision for the GIS?
- What is the goal for GIS technology?
- What are the short- and long-term goals and objectives?

STRATEGIC

COMPONENTS

What is the vision for GIS? • What is the goal for GIS technology? • What are the short-term and long-term goals and objectives? • How do we define our goals and objectives? • How will GIS enhance our functions? • What are the priorities for services and GIS functions? • How will our vision align with the overall goals and objectives of the organization? • What type of governance model should be used? • What hurdles might we encounter? • How can we best use intergovernmental agreements? • How can we best use Data Sharing Agreements? • Do we need to use Service Level Agreements within the organization? • What about the regionalization of GUS and how does it fit into the vision? • What role does the GIS Steering Committee have in the governance of GIS? • Do we have bottom up and top down support for enterprise GIS? • How will we promote our solution? • How will we train and educate staff?

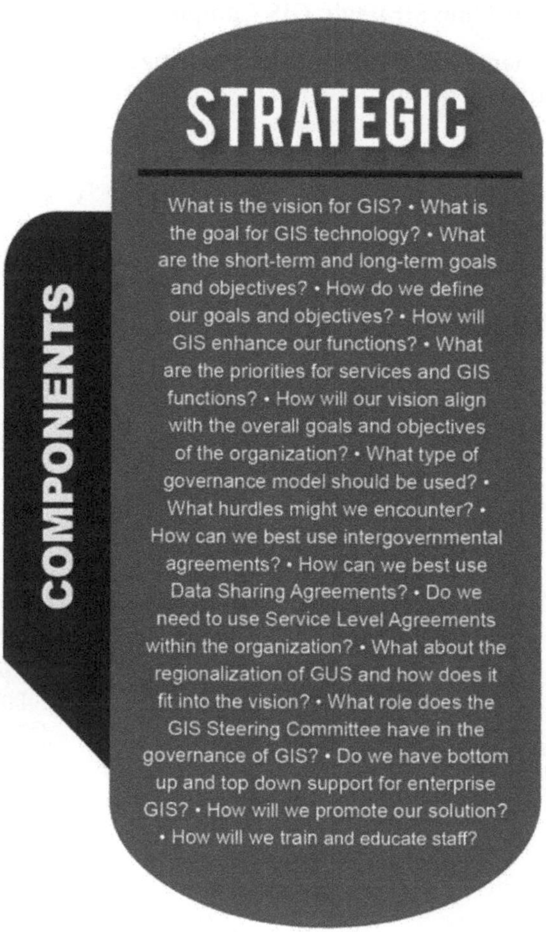

FIGURE 4.6
Strategic components.

- How do we define our organization's goals and objectives?
- How will the GIS enhance our organization's functions?
- What are the priorities for services and GIS functions?
- How will our vision align with the overall goals and objectives of the organization?
- What type of governance model should be used?
- What hurdles might our organization encounter?
- How can our organization best use intergovernmental agreements?
- How can our organization best use data-sharing agreements (DSAs)?
- Do we need to use SLAs within our organization?
- What about the regionalization of GIS, and how does it fit into vision?
- What role does the GIS steering committee have in the governance of GIS?
- Do we have bottom-up and top-down support for enterprise GIS?
- How will we promote our solution?
- How will we train and educate our staff?

4.3.2 Component Two: Tactical GIS Components

Tactical components are the things that are considered, prepared, designed, and scheduled to meet the organization's goals and objectives. They are the details within the larger strategy. Boots-on-the-ground tactical planning solves operational problems and answers the questions in Figure 4.7.

Figure 4.7 is a list of the tactical components of an enterprise GIS.

- How will our organization manage the GIS?
- How will we enforce the governance model?
- What type of GIS users should exist within the organization?
- What general policies and procedures are needed?
- How best should we use a GIS software licensing model?
- How can we best utilize local, state, and federal data?
- How should we use intergovernmental agreements and DSAs?
- How do we implement and enforce SLAs and DSAs?
- Are there any opportunities to generate revenue from the enterprise GIS. How do we deploy a cost recovery strategy?
- How do we prioritize GIS integration with all other systems?
- What conferences should staff attend?
- Do we need a GIS user group?

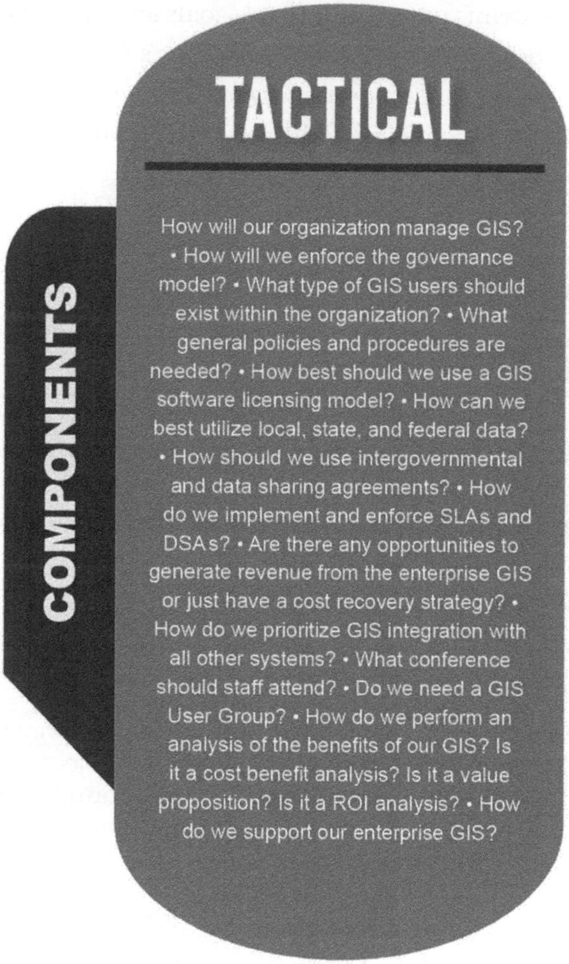

FIGURE 4.7
Tactical components.

- How do we perform an analysis of the benefits of our GIS? Is it a cost–benefit analysis? Is it a value proposition? Is it an RoI analysis?
- How do we support our enterprise GIS?

4.3.3 Component Three: Technical GIS Components

The technical components include the actual mechanical, scientific, procedural, and specialized parts of the GIS implementation process. The technical components include the software technology, the IT infrastructure, and all the management bits in between. The proverbial *devil in the details*

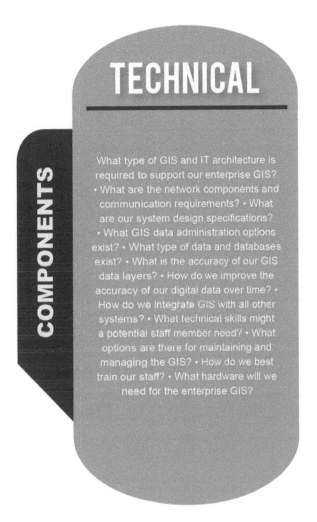

FIGURE 4.8
Technical components.

can be addressed by answering the following questions that are listed in Figure 4.8.

- What type of GIS and IT architecture is required to support our enterprise GIS?
- What are the network components and communication requirements?
- What are our system design specifications?
- What GIS data administration options exist?
- What type of data and databases exist?

- What is the accuracy of our GIS data layers?
- How do we improve the accuracy of our digital data over time?
- How do we integrate the GIS with all other systems?
- What technical skills might a potential staff member need?
- What options are there for maintaining and managing the GIS?
- How do we best train our staff?
- What hardware will we need for the enterprise GIS?

4.3.4 Component Four: Logistical GIS Components

Logistical components comprise everything that is related to the rational supply and support of the technical tools and procedures. Figure 4.9 is a list of the logistical components of an enterprise GIS. These components should answer the following questions:

- Who should perform certain and specific GIS functions?
- Who manages all the components of a GIS?
- What staff support and contractual services are needed?
- Is it possible for existing staff to perform GIS work?
- What are the (ongoing) costs of GIS implementation?
- What budget model do we use?
- Could our organization's resources better support the GIS?
- How do we educate our staff?
- What tools do we have to transfer knowledge?
- Do we have succession planning (strength in depth)?
- How do we monitor and measure user satisfaction?
- How do we track support of the GIS?

4.3.5 Component Five: Political GIS Components

Political components are the governmental and ethical components of GIS implementation. Figure 4.10 shows a list of the political components of an enterprise GIS. These include considering the needs of elected officials and the mission of the organization and answering the following questions:

- How do we sell GIS to decision makers?
- How do we secure buy-in from elected officials?
- Is our GIS aligned with the organization's vision, goals, and objectives?
- How do we address ethics and GIS?

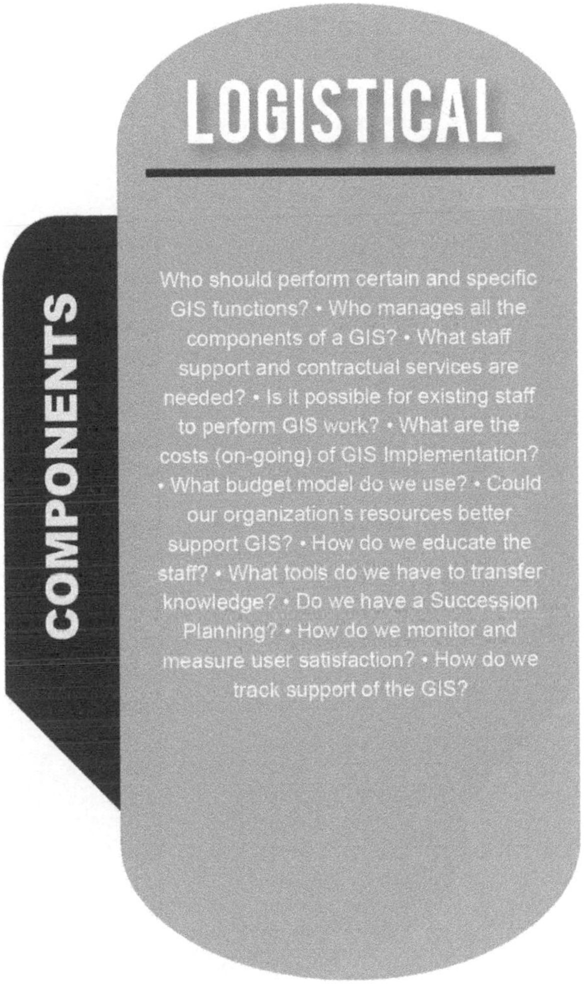

LOGISTICAL

COMPONENTS

Who should perform certain and specific GIS functions? • Who manages all the components of a GIS? • What staff support and contractual services are needed? • Is it possible for existing staff to perform GIS work? • What are the costs (on-going) of GIS Implementation? • What budget model do we use? • Could our organization's resources better support GIS? • How do we educate the staff? • What tools do we have to transfer knowledge? • Do we have a Succession Planning? • How do we monitor and measure user satisfaction? • How do we track support of the GIS?

FIGURE 4.9
Logistical components.

- What legal issues need to be addressed?
- How do we enable decision makers?
- How do we measure and quantify results from the GIS initiative?
- How do we present GIS successes?
- How do we create intergovernmental agreements and DSAs?
- What do we do about the regionalization of GIS?
- How do we secure funding for the GIS?
- What is the value of GIS?
- What about cost recovery versus revenue generation?

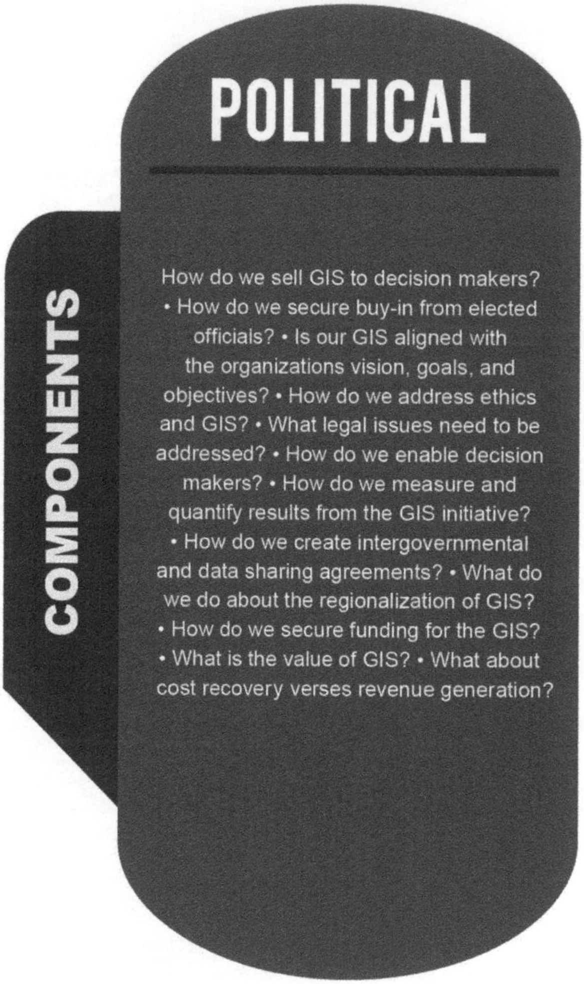

FIGURE 4.10
Political components.

4.4 Challenges, Barriers, and Pitfalls

Over the past 40 years, we have seen a widespread adoption of geospatial technologies in local government. Periodicals are cover to cover with success stories. Esri has grown in leaps and bounds, and geospatial technology is a massive-growth industry. However, I have evidence that suggests that GIS technology is still significantly underutilized in towns, cities, and counties across the United States. I gathered this evidence through my firsthand knowledge of working with hundreds of organizations, extensive benchmarking of municipalities, literature search, and conversations with local government professionals. The major challenges, barriers, and pitfalls are associated not only with the implementation of a true enterprise GIS in local government but also its ongoing operation and end-game success. I am a big believer in compartmentalizing. Things become manageable when broken down, so let us discuss what we mean by challenges, barriers, and pitfalls and examine what solutions exist to remedy these constraints. The following details in Figure 4.11 are the most common challenges, barriers, and pitfalls of GIS technology.

4.4.1 Challenges to an Enterprise GIS

The challenges to deploying enterprise GIS are often strategic and tactical in nature. They tend to include issues that are associated with the implicit demand for proof, an objection, a query, a test, or a dispute about the validity of GIS. Figure 4.12 lists the challenges. Challenges include:

- *People*—Securing buy-in for the big picture is critical. Challenges can come from reluctant senior staff.
- *No strategic plan*—You must develop a multiyear GIS strategic implementation plan that directs growth and secures the budget.
- *Lack of a GIS vision*—You must create a vision for your organization. This is part of the planning process.
- *No goals or objectives*—Very specific departmental goals and objectives are essential.
- *No quantifiable return on investment (RoI)*—You must be able to present the tangible and intangible benefits and costs of GIS and discuss the RoI.
- *Not selling or promoting GIS success*—Every organization requires talented staff to sell GIS technology and promote its virtue.
- *No GIS alignment*—Aligning the GIS with the mission of the organization is often overlooked.
- *The funding mechanism*—You must always follow the money. Make sure that the GIS is funded through reliable sources.
- *Poor enterprise data accuracy and lack of an understanding of data quality*—Incomplete and inaccurate digital data can jeopardize the GIS initiative. A comprehensive digital data assessment is always a challenge.
- *No endorsement from elected officials*—You must secure buy-in from not only the officials and management but also the elected officials who oversee the organization.
- *Lack of GIS regionalization*—A lack of cooperation and coordination with other organizations could prevent success.
- *Poor enterprise project management*—Managing and coordinating the intradepartmental GIS is critical to the success of the organization.

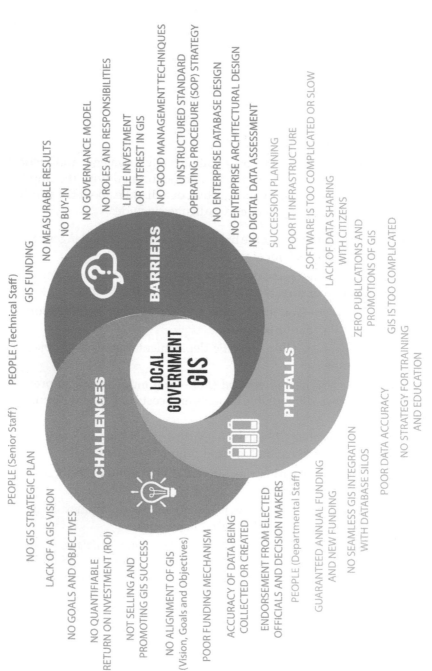

FIGURE 4.11
Challenges, barriers, and pitfalls.

CHALLENGES

☑ People
☑ No Strategic Plan
☑ Lack of a GIS Vision
☑ No Goals and Objectives
☑ No Quatifiable Return On Investment (ROI)
☑ Not Selling or Promoting GIS Success
☑ No GIS Alignment
☑ The Fundings Mechanism
☑ Poor Enterprise Data Accuracy and Lack of Data Quality
   Understanding
☑ No Endorsement from Elected officials
☑ Lack of Regionalizations of GIS
☑ Poor Enterprise Project Management

FIGURE 4.12
Challenges.

4.4.2 Barriers to an Enterprise GIS

The barriers to deploying enterprise GIS tend to be technical and logistical in nature. They are the obstacles that the project faces, the hurdles, the stumbling blocks, the hindrances, and the impediments. They keep people or processes apart and prevent communication or progress. Figure 4.13 lists some barriers. Barriers include:

- *People*—Communication, knowledge transfer, and support are a key element.
- *GIS funding–departmental payback model*—Funding models stifle growth.
- *No measurable results*—You must show measurable and attainable successes.
- *No buy-in from staff*—You have to secure buy-in from the *rank-and-file* members of an organization.

BARRIERS

☑ People
☑ GIS Funding - Departmental Payback Model
☑ No Measurable Results
☑ No Buy-In from Staff
☑ No Governance Model
☑ No Roles and Responsibilities
☑ Little Investment
☑ No Good Management Techniques
☑ Unstructured Standard Operating Procedures
☑ No Enterprise Database Design
☑ No Enterprise Architectural design
☑ Deployment of Enterprise Software

FIGURE 4.13
Barriers.

- *No governance model*—You must understand your governance model in order to improve the management and coordination of your GIS.

- *No roles and responsibilities*—Clearly defined GIS roles and responsibilities and accountability are key to success.

- *Little investment*—You must make good GIS investment decisions. The GIS budget is very important. Some departments with local government have enterprise funds. Some departments do not have funds to invest but actually need GIS. The parks and recreation department is generally an example.

- *No good management techniques*—Remove the obstacles to success from the top down.

- *Unstructured standard operating procedures*—It is very important to have repeatable and understandable standards and protocols.

- *No enterprise database design*—The design of an enterprise database is critical to success.

- *No enterprise architectural design*—An enterprise architecture that addresses hardware, software, and communication will play a significant role in success.

- *Deployment of enterprise software*—Web-based tools, along with Intranet and Internet solutions, are key to success.

4.4.3 Pitfalls to an Enterprise GIS

Pitfalls refer to the hazards along the way. They refer to a hidden or unsuspected difficulty in the implementation process. Think of a snare, a trap, or a maze. Figure 4.14 lists the pitfalls.

- *People*—Succession planning or GIS strength in depth often plays a key role and can hamstring an organization if improperly handled.

PITFALLS

☑ People
☑ No Guaranteed Annual Funding
☑ No Seamless Integration with Database Silos
☑ Poor Departmental Data/Accuracy/Duplication
☑ No Strategy for Education and Training
☑ GIS is too Complicated
☑ Zero Publications and Promotions
☑ Lack of Data Sharing with Citizens
☑ Lack of Availability of GIS Software
☑ Software is too Slow
☑ Poor IT Infrastructure
☑ Lack of Succession Planning

FIGURE 4.14
Pitfalls.

- *No guaranteed annual funding*—You must identify funding and investment strategies.
- *No seamless integration with database silos*—You must optimize and maximize the use of GIS for efficiency and promotion.
- *Poor departmental data accuracy/duplication*—You must make your data accurate. Eliminate data inaccuracies, redundancies, and duplication.
- *No strategy for education and training*—You must have an education plan. Educate and inform your organization regularly.
- *GIS is too complicated*—Make GIS easy to use and indispensable.
- *Zero publications and promotions*—Promoting the effectiveness of GIS is crucial to ongoing success.
- *Lack of data sharing with citizens*—Citizen engagement is a component of enterprise GIS that goes a long way toward securing further funding.
- *Lack of availability of GIS software*—Restricting GIS use and difficult-to-access GIS software will prevent success.
- *Software is too slow*—Up-to-date software optimizes user interaction.
- *Poor IT infrastructure*—Firm IT infrastructure is necessary for supporting such a massive, digital-age project like GIS implementation.
- *Lack of succession planning*—Developing expertise and internal depth is a key for an ongoing enterprise solution.

We now have a complete list of all of the challenges, barriers, and pitfalls of any enterprise, sustainable, and enduring GIS. The next two questions to answer include, (1) which GIS challenges, barriers, and pitfalls are most common in local government, and (2) when in the process of GIS maturation do these constraints appear?

4.5 The Challenges, Barriers, and Pitfalls of 100 Organizations

I have worked with hundreds of local government organizations to plan, design, and implement GIS solutions. During the writing of this book, I decided that it would be useful to summarize the challenges, barriers, and pitfalls of 100 organizations that I have encountered. During the development of GIS strategic implementation plans that included Blue Sky think tank sessions, interviews, and questionnaires, each organization noted specific challenges, barriers, or pitfalls. They were essentially documenting the reasons that prevented GIS from being successful. This is not a very scientific analysis, but it does give us an insight into the most common and most influential constraints. Figure 4.15 represents my findings. This figure illustrates the percentage of local governments that thought that a specific challenge, barrier, or pitfall was the problem.

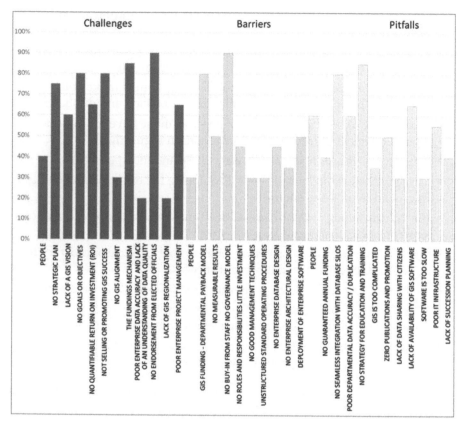

FIGURE 4.15
GIS constraints of 100 local government organizations.

4.6 The Challenges, Barriers, and Pitfalls during GIS Maturation

Figure 4.16 illustrates GIS maturation or three GIS development stages. It explicitly illustrates when challenges, barriers, and pitfalls may be encountered in an immature, mature, or advanced organization. It's the challenges along the way you have to be prepared to address. So, if you are an organization just starting off in GIS, be prepared to develop a plan of action, identify funding to support your enterprise GIS, develop a vision and goals and objectives, build reliable and repeatable digital data, and arm yourself with the knowledge that you will have to continually sell GIS to your organizations.

If you are an organization that is entering the mature stage of GIS deployment, forewarned is forearmed, consider Standard Operating Procedures (SOPs), focus on creating and improving accurate digital data, develop and

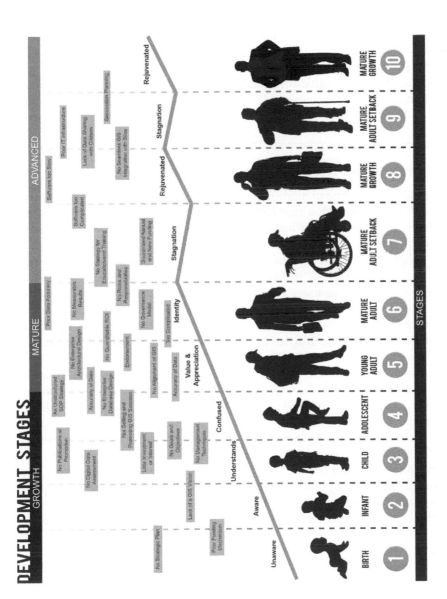

FIGURE 4.16
GIS maturation in local government.

deploy an enterprise architecture, address good governance and roles and responsibilities, and most importantly understand GIS lines of communication. It is at this stage where it is imperative that you align your GIS vision with your organization's overall goals and objectives.

If your organization is in the advanced stage of GIS, you will be addressing data standards, software functionality and ease of use, funding models, ongoing training, education and knowledge transfer, regionalization of GIS, intergovernmental agreements, IT infrastructure, succession planning, and last but not least, the politics of GIS. Aligning and selling GIS still play a significant role in your success.

4.7　The Role and Responsibilities of the GIS Coordinator

I hope that by now, I have made it clear that there are many components that are involved in building a true enterprise solution. By its very nature, there is also an abundance of challenges, barriers, and pitfalls that will ultimately prevent an organization from succeeding. Now, I would like to discuss why some people think that the number one reason for failure to implement a successful and enterprise, sustainable, and enduring GIS is the inability of the GIS coordinator to meet all implementation requirements. I definitely think that all roads lead directly back to the GIS coordinator, but this is only part of the story. You could say that if a GIS coordinator cannot perform all of the following enterprise GIS duties that are documented in Figure 7.11, then the GIS will be less than successful.

What would you do if I told you that you are ill-equipped to manage an enterprise, sustainable, and enduring GIS? What if I told you that your management style is not working? What if I told you that you are doing the job of five people, and it will take talent, dedication, and a formula, or at least an understandable playbook, to succeed? Are you still reading?

At the start of this chapter, do you remember me asking the question, is the number one reason for failure to implement and maintain an enterprise, sustainable, and enduring GIS the failings of the GIS coordinator? The hypothesis is that all of the challenges, constraints, and barriers would go away if an experienced and talented GIS coordinator was running the show. If we play the *blame game*, we would blame all ills on the GIS coordinator.

There is enough evidence to illustrate the extraordinary ability of local government professionals to demonstrate human creativity. The annual Esri Local Government Awards illustrates the variety and range of this creativity. I believe that local government professionals and the local government arena are two of the greatest forums for making the world a better place. An individual or organized group of people in local government with passion and armed with an incredible, invaluable, and intrinsic incentive can make more impact on our world than any financially incentivized business operations. The capacity for creativity and innovation exists within our local government.

Let us look at the business world and constraints that prevent success. It is argued that the top five factors for success in a startup business include the following:

1. Timing
2. Team and execution
3. Idea
4. Business model
5. Funding

This is according to Bill Gross' 2015 TED talk. Let me ask you a question. What is the art of good comedy? The answer is timing. It is only funny if you say it quickly after the question is asked. In the business world, *timing* seems to be an incredibly important, if not the single most important, factor determining the success of a startup business and solution. I think that there is enough evidence to support timing as the most important factor. The second factor is the team, and then the actual idea, followed by the business model, and last, but not the least, is the funding. Other professionals include communication, originality, and value as factors that prevent success. So where does this leave us with the GIS in local government? There seems to be some overlapping. Even though what I am about to say contradicts the findings that are illustrated in Figure 4.15, I think that the top five priority challenges, barriers, or pitfalls in local government include the following:

1. No strategic GIS plan
2. Poor governance
3. No vision, goals, objectives, and alignment
4. No enterprise software
5. Unreliable funding

A GIS coordinator has to possess an incredible amount of organizational skills, imagination, flexibility, original thought, problem-solving, and social agility to coordinate and collaborate with elected officials, senior management, and departmental staff. He or she has to be a great communicator and presenter; be an educator; have the ability to delegate; understand the art of cooperation; inspire a shared vision, goals, and objectives; and possess extraordinary team-building skills.

What I am saying is that local government GIS coordinators need to leap tall buildings in a single bound. Figure 4.17 gives a simple, graphic-rich illustration of all of the components that we have addressed in this chapter.

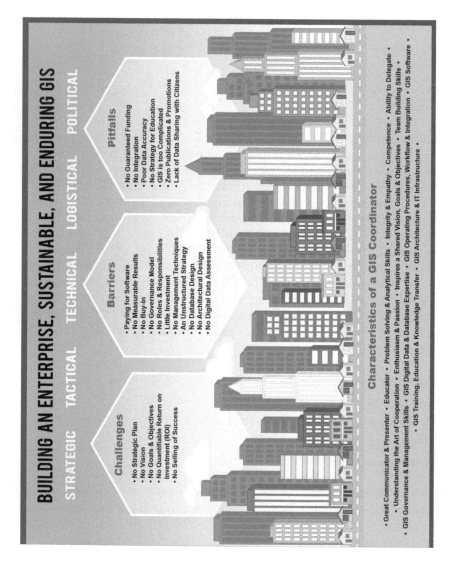

FIGURE 4.17
Overview of challenges, barriers, and pitfalls.

5

Sequential Steps to Developing a Vision, Goals, and Objectives

In the long history of humankind those who learned to collaborate and improve most effectively have prevailed.

Charles Darwin

5.1 Introduction

Never underestimate the potential outcomes of a ratified GIS vision statement that is supported by specific goals and objectives. I do not think that anyone, anywhere would challenge the idea that having a vision for the future of your organization's GIS is necessary. There are hundreds of books, blogs, Websites, and tweets that emphasize the importance of having a mission statement, a vision statement, goals, and objectives. GIS is no different from any other initiative that requires planning, foresight, and vision. Ronald Reagan was quoted saying, "To grasp and hold a vision is the very essence of successful leadership." And I support the notion that belief and leadership play a distinct role in the success of an enterprise GIS.

Most organizations lacking a GIS vision usually fail at implementing an enterprise solution. Even those organizations that do have a vision statement often fail because they do not follow the steps that they themselves laid out for success. GIS implementation is not an overnight scuffle. It is a multiyear battle. A clear vision, supported by the necessary steps and the energy, belief, and enthusiasm to complete the tasks outlined, is a critical component of an enterprise GIS. Do not lose sight of your goals. Avoid distractions from completing those step-by-step action plans.

Developing a vision should be an organic process and culturally significant. It should be creative and analytical. It should involve focused discussions about the experiences, ideas, and the perceived function of the organization. It was rumored that Mark Twain said, "Twenty years from now you will be more disappointed by the things you didn't do than by the ones you did do. So throw off the bowlines. Sail away from the safe harbor. Catch the trade winds in your sails. Explore. Dream. Discover." Take these words to heart when arranging your vision for implementing a GIS initiative.

In the interest of making this chapter understandable, I developed a linear approach to developing a vision, goals, and objectives for your organization. A crystal-clear vision statement that details the steps necessary for an organization to achieve its vision should be your primary goal as a GIS coordinator. After all, if you succeed in accomplishing this task, everything else you are charged with should fall into place.

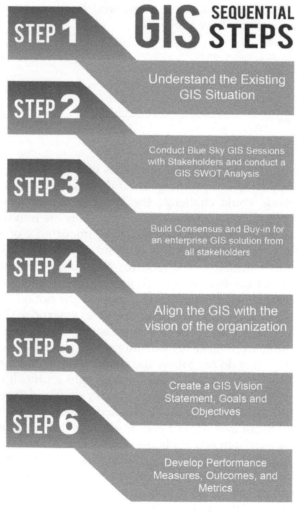

FIGURE 5.1
Sequential steps to developing a GIS vision, goals, and objectives.

To successfully develop a vision, goals, objectives, and supporting performance measures, you should organize your thought around the following six steps that are illustrated in Figure 5.1.

1. Step one: Understand the existing GIS situation
2. Step two: Conduct Blue Sky GIS sessions with stakeholders and conduct a GIS strengths, weaknesses, opportunities, and threats (SWOT) analysis
3. Step three: Build consensus and buy-in for an enterprise GIS solution from all stakeholders
4. Step four: Align the GIS with the vision of the organization
5. Step five: Create a GIS vision statement, goals, and objectives
6. Step six: Develop performance measures, outcomes, and metrics

5.2 Step One: Understand the Existing GIS Situation

It is both impossible and unwise to implement lasting strategic changes without the consent and cooperation of the majority of an organizations' staff. Thus, the first step in this model is to gather as much information of the existing situation. Chapter 2 details the recommended strategy for data gathering, and Chapter 3 outlines all of the components that are necessary to deploy an enterprise solution. That withstanding, we need to conduct an unbiased, objective review of the organization's current GIS capabilities, and resources, as seen through the eyes of all staff. Everything that is linked to the existing GIS initiative must be documented and detailed including the following:

- *Governance*—A detailed description of the existing governance model
- *Digital data and databases*—An assessment of the accuracy and reliability of GIS data
- *Procedures, workflow, and integration*—Existing procedures, workflow, and integration
- *Software*
  - Number of users
  - Types of users
  - Software licensing

- *Training, education, and knowledge transfer*—Existing activity
- *Infrastructure*—Hardware, infrastructure, and communications

Following an analysis of the existing situation, results can be easily packaged and presented. Step one is just about detailing the current conditions, less so much about people's feelings, ideas, and thoughts about the future. That comes in step two. Figure 5.2 shows the findings of an existing conditions assessment.

5.3 Step Two: Conduct Blue Sky GIS Sessions with Stakeholders and Conduct a GIS SWOT Analysis

Blue Sky sessions are brainstorming sessions where all staff are allowed to speak freely and openly. The Blue Sky sessions I have moderated entailed an open discussion with a relatively small group of people regarding new ways to understand their existing situation and provide solutions to the GIS-related problems of the organization.

I believe in blending Blue Sky sessions with a classic SWOT analysis. Frankly, formal meetings *do not* produce the results that are gathered in Blue Sky and SWOT sessions, though these sessions do require a skilled moderator to facilitate the information gathering. It is critical during these processes that ideas are filtered so that the unworkable, plain-wrong, and ill-formed notions are weeded out early on. Brainstorming sessions should be a multistage process and entail the following:

- Ideas and thoughts about the organization, especially those that are related to the SWOT analysis
- Combining and refining the ideas that are gathered
- Deliberation on and selection of valid ideas and observations
- Creation of a SWOT list that is supported by innovative ideas

Figure 5.3 illustrates the preliminary findings of a SWOT analysis. A SWOT analysis is a strategic planning method that is used to evaluate the strengths, weaknesses, opportunities, and threats as they relate to all components of GIS implementation. After the GIS needs assessment is complete, a GIS SWOT analysis is performed on-site. This allows staff to understand the purpose of the strategic planning process.

Statement of Findings

GOVERNANCE	DATA & DATABASES	SOFTWARE	TRAINING
Each department participated.	The majority of people do not edit - » data layers » custodianship issues » enterprise issues	The majority of people say they use the enterprise GIS.	The majority of **people have a knowledge of what GIS can do.**
Most people believe that GIS can benefit them.		The majority of city staff use **web applications** for GIS use.	The **majority of people spend 1-25% of their time using GIS.**
Most people believe that GIS will - » Improve data accuracy » Improve efficiency » Improve information processing	Most people are not sure about clear lines of responsibility regarding data creation/maintenance.	The majority of people believe GIS is difficult to use.	Most people have **received training.**
Most people see **data viewing** as the primary GIS activity.	The most important "technical needs" are - » data standards » digital submissions » meta data » data documentation	Most people do not use GIS on a daily basis.	The most important **"logistical need"** is **training and education.**
Most people believe that the existing GIS is effective at meeting the needs of their department.		Most people use map browsers.	Most people believe that GIS **training and education is an important component** of a true enterprise solution.
The most important **"strategic need"** is departmental goals and objectives.	Most people consider the areas of **"mapping and access to GIS data to be areas of high priority."**	The most important **"tactical need" is new uses of GIS.**	Most people **desire more GIS knowledge.**
The majority of people go to another person in their department for their GIS/mapping needs.	**Key data layers are incomplete and inaccurate.**	Some users are frustrated by software crashing or slow access to data.	Most people would **use online training services.**
		A number of people expressed an interest in **mobile GIS.**	Most people agree that any successful GIS would include - **training, education, and knowledge transfer.**

FIGURE 5.2
Existing GIS conditions assessment.

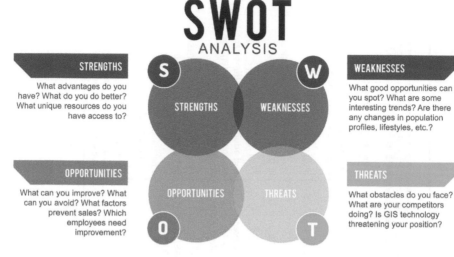

FIGURE 5.3
Blue Sky SWOT analysis and findings.

For example, a city in Canada set up functional SWOT teams. The following, as illustrated in Figure 5.4, were their functional GIS SWOT groups:

- *Community information and communications*—Staff who focus on interacting and sharing data with the public
- *GIS editors*—Staff who are responsible for the creation and editing of GIS data, using advanced GIS tools
- *GIS user group*—Members of the existing GIS user group
- *Mobile and field*—Staff representing departments/divisions that have a pressing need to use the GIS in the field
- *Property and address*—Staff who focus on managing property and address-related data
- *Public safety*—Staff whose primary duty pertains to public safety and welfare
- *Steering committee*—The existing GIS steering committee
- *Utilities*—Staff from departments that are responsible for infrastructure and utilities

The results of the SWOT analysis were utilized to make recommendations for advancing the enterprise GIS. In summation, the SWOT analysis identified the following:

- *Strengths*—Characteristics that placed the city in an advantage for implementing an enterprise-wide GIS.

Community Information & Communications
Staff who focus on public interaction and sharing of data with the public

Steering Committee
The existing GIS Steering Committee

GIS Editors
Staff who are responsible for the creation and editing of GIS data with advanced GIS tools

GIS User Group
The existing GIS User's Group

FUNCTIONAL
SWOT
TEAMS

Mobile & Field
Staff representing departments/divisions that have a pressing need to use GIS in the field

Property & Address
Staff who focus on managing property and address related data

Utilities
Staff from departments focusing on infrastructure and utilities

Public Safety
Staff whose primary job duty is public safety and welfare

FIGURE 5.4
Functional SWOT teams.

- Enthusiasm for the technology, a desire to improve organizational operations, and a clear set of identified departmental needs provided fertile soil for advancing the GIS enterprise-wide.
- Previous successes laid the groundwork for future successes, including the following:
  - Implementation of Esri software
  - Culture map
  - Integration with existing systems
  - Field collection for trees, road patrol, and environmental items
  - Mobile pilot project
  - Intranet-based GIS viewer

- Expert-level staff exists within the city government. A number of departments employ adept GIS staff who are prepared to lead the city in advancing the GIS.
- Leadership—The city formed a GIS steering committee and user group in an effort to direct GIS efforts.
- The city already held a wealth of existing data.

Figure 5.5 documents the strengths of the organization.

- *Weaknesses*—Challenges or weaknesses related to the implementation of an enterprise-wide GIS.
  - Lack of a clearly defined GIS governance structure.
  - Financial constraints.
  - Competing priorities—the city has many large projects ongoing.
  - Lack of training and education.
  - Data inaccuracies—attribution, position, and temporal.
  - No uniform mobile solution.
  - Speed and performance of existing tools.

Figure 5.6 documents the weaknesses of the organization.

- *Opportunities*—Issues related to improved organizational effectiveness and efficiency.
  - Public access and public involvement.
  - Formalized governance that enables all users (internal and external).
  - GIS integration with existing systems (adding value to those systems).
  - Return-on-investment (RoI) impact (ice storm, green initiative).
  - Easy-to-use tools.
  - Using GIS in the field (mobility).
  - RoI.

Figure 5.7 documents the opportunities of the organization.

- *Threats*—Issues the city may encounter that threaten an enterprise-wide GIS solution.
  - Provincial and federal mandates slowed data dissemination.
  - Reluctance to change or a cultural shift.
  - For budgeting, the city must justify spending via RoI possibilities.

STRENGTHS

Community Information and Communications	GIS Editors	GIS User Group	Mobile and Field	Property and Address	Public Safety	Steering Committee	Utilities
» Open records for all organizations » Open to "any" GIS » Online browser accessibility - Software as a Service » Culture Map » Desire » Class software/data » Integration with GIS	» Software (Esri) » Training	» Integration (Asset Management Work Order) » Data (Abundance) » Enterprise Use » Good IT Department » Development of Relationships (Strategic Project Successes, External Organizations) » Enthusiasm from all levels » Recognition of Values GIS (ROI)	» Existing ArcPad Solutions » Data (Mobile Ready) » Species Collection Success Story » Automated Data to GIS for Storm Water field staff » Identifying Road issues in the field - Pot holes » Governance - Available and User friendly GIS staff in IT » Hardware (Connection Citrix, Amada, inspections)	» Political/Admin will support » Data Layers (Abundance) » On-point (Citizen and Departments) » Business Solutions » Willingness to Try New Technology - ArcGIS Online » Ground swell of support from staff » Steering Committee in Place - Recognize need for more governance » Positive	» EMS Open to GIS » Many Needs and Desires » Request for funds for Analysis location of station - understands real ROI possibilities » C&D used GIOS on a daily basis » Improve response times because of GIS » Good Data (City) » Good GIS Staff	» Recognize value of Education, Training and Knowledge » Good City Products and Services » History of Success » Existing Governance » Drive to use Technology » Plans-Existing Strategies from Council (GIS can assist with these) » Existing Steering Committee » Skilled GIS Staff for Corporate Style GIS » Existing Investment in Technology (Esri)	» Dedicated staff in Utilities (Water/Engineering) » Accurate Date » Infrastructure Asset Management System » Other systems use GIS ID's » Cross section of data layers » Responsible Use of GIS » GIS Trained » Understand the value of shared GIS data

FIGURE 5.5
SWOT strengths.

WEAKNESSES

Community Information and Communications	GIS Editors	GIS User Group	Mobile and Field	Property and Address	Public Safety	Steering Committee	Utilities
» Lack of Governance » Resources » Communication » Training » Education » Data (Accuracy, Not exist, Not useable, Understandable, What type of data) » Staff Capacity » Budget » Structured » Dedicated » No GIS Program » Sustainability	» Data Standards » No formal governance structure » Vulnerable to staff leaving » SOP » Infrastructure in a pickle if staff leaves » It Network Speed » Duplication of Data Layers » Lack of Training (Departmental) » Not Centralized in regards to data » Budget-Communication » Lack of AutoCAD standards » Corporate Reliance on GIS Technical Editors	» Lack of Mobile GIS » Funding » Lack of Education and Training » Data Lack of Centralization » No Metadata » Lack of Understanding » Integration-Enterprise » Documentation (SOP) » Technical Support | » Governance » Lack of Resources	» Need a Mobile Specialist in IT to understand » IT Infrastructure » Coverage of Cell Phones » Governance » 5 Groups using Mobile now - not Uniform » No Uniform Strategy » GPS Tools Sustainability » Accuracy of Data » Meta Data » Communication of Technology (Cell Coverage) » Missed Opportunity	» Funding » Duplication of Effort » No Clear Understanding –Department Specific » The Explanation of ROI » Ongoing Education » Corporate Structure » Save Cost-Not Funded » Culture? » Buy-in » Data Accuracy » Data Maintenance issues » Clear lines of Responsibility	» No skill set or training » Dependent upon GIS staff » Access to GIS » Reliant on Provence and City (IT Infrastructure) » Lack of Uniform System » No Optimal Reserves (Funding) » Operational Weakness? » Where are vehicles	» Training and Education Funding » Problems with Multiple Initiatives » Lack of Corporate View of Systems » Need Expert GIS Knowledge (WAM) » Dedicated Funding » Lack of Corporate GIS Knowledge – what can GIS do for me and my staff » No GIS Lead » Lack of formal Governance Structure » Lack of Meta Data » Lack of Standards	» Need Local Government Information Model » Data accuracy Gaps (Easement Layer) » Attitude towards GIS - History of Personnel who did not support » Lack of Accuracy - Positional » Lack of Field Access » Lack of SOP for Data Entry/Collection » Lack of Governance » IT Infrastructure – too slow (10MBPS pipe) » Not Uniformly Engaged » DPI (Divisions different places in GIS expertise » Lack of Geometric Models - Intelligent Networks

FIGURE 5.6
SWOT weaknesses.

OPPORTUNITIES

Community Information and Communications	GIS Editors	GIS User Group	Mobile and Field	Property and Address	Public Safety	Steering Committee	Utilities
» Public Involvement » Citizen Engagement » GIS Centric Tool (Visual Techniques) » Business Analytical Tool (Build Case-GIS tool kit) » Open Data Team » Open Data 'Day' » Open Transparent Government, Open Government	» Corporate Style GIS » Centralized of Data » Data Standards » Formalize Governance » Documentation of data and process » Data Warehouse » Define Roles and Responsibilities » Formal Business Process (SOP) (Work Flow, Integration with systems) » Technology- Have a common Corporate Solution and Uniform Strategy	» Routing-Road Closures » Applications of GIS » Data Proactive Citizen Notification » Collaboration » Open Government-Transparent Government » Citizen GIS 'How can we help' 311 » Location based GIS » Excitement and Enthusiasm	» Improved Efficiency » Field Use - better decisions » Paperless City » Willingness to use Tools » Response to Events (Ice Storm) » Understanding Digital Data » Citizen Involvement (311) » Better Analysis When Needed » Uniform Solution (Support Structure, Reduce Litigation Issues) » Applications- Arc GIS Online – utilize ESRI	» Mobility of GIS » Corporate Collaboration » Additional Digital Layers » Training/Education » Improve IT/Speed » Confidence in Data » Public/Citizen Engagement » Increase in staff productivity	» Access to data » In-house Study –station location study » New AVL system » RFP to replace GREY Island » Mobility of GIS » Education » Training » Knowledge » Common Operational Picture (COP) GIS » More Data » Geo Enabled Systems » Geo references	» Political Opportunities (Ice Storm, Green Initiative Need Education) » Build on Existing Plans (Water, Energy, IOR, Forestry, Open Government, Storm water, Community, Parks, Source Water) » Perfect time to 'Embrace' GIS other initiatives » Public Awareness » Public Participation 311	» Need Software Integration- Speak to each other » Need GIS Centric Asset Management & Work Order Solution » Use GIS for Sewer (All uses- it's not used) » Visually see work & Financial data » Public Communications » Health and Safety Lone Worker uses » Educate and Training 'Knowledge Gap'

FIGURE 5.7
SWOT opportunities.

- Failure to implement an enterprise-wide GIS governance structure.
- A too technical and complex system results when a city overlooks the importance of getting the right tools into the hands of the users.
- Focusing too much on other initiatives at the expense of GIS.
- Erroneous data that undermined the validity of GIS.
- Lack of integration with other systems (stovepipes).

Figure 5.8 documents the threats to the organization.

5.4 Step Three: Build Consensus and Buy-In for an Enterprise GIS Solution from All Stakeholders

The people of an organization are the core foundation for a successful GIS. Thus, *consensus building* or *consensus decision making* is fundamental for successfully developing an organization's GIS vision, goals, and objectives. They say involve others to increase commitment and engagement. I do not really know who *they* are, but they also say that a staff person who has a hand in creating its own vision, goals, and objectives is more likely to succeed.

Consensus describes a collective agreement to support an idea or decision that is in the best interest of the organization. Consensus is founded on mutual understanding, an agreement to support a decision, and a commitment to take active steps for the benefit of the group. A consensus decision does not imply that everyone agrees on all of the details, or that individuals changed their perspectives or opinions on a situation. Consensus decision making is a process that builds trust and creates ownership and commitment. An effective consensus process (consensus building) is inclusive and engages all participants. Consensus decision making leads to outcomes that are not only higher in quality but would also empower the GIS group or community to move forward and create a future together. Consensus-building characteristics include the following:

- Inclusive participation.
- Engagement and empowerment of the group.
- A commitment to work together.
- Increased cooperation.
- Shared understanding.
- Bridging of differences.
- Equalization.
- Higher-quality decisions.

THREATS

Community Information and Communications	GIS Editors	GIS User Group	Mobile and Field	Property and Address	Public Safety	Steering Committee	Utilities
» AODA Compliance » Staff Capacity » Corporate Culture-Reluctance to Change (Compliance, Freedom of Info, Open Shared Data, Open Legislation) » Privacy (Accessibility to Data) - Must understand "off limits" data » Election » Buy-in » ROI Business case » Budget-How much will it Cost vs ROI » Exclusion of Citizen Technology	» Lack of Understanding » Budget » Governance » Training » Staff Turnover » Lack of Communication » Staff Time (Zero time for metadata) » Citizen Expectations » More Data = Slower IT	» Resistance to Change » Corporate Support (Buy-in) » Too Technical or Complex » Organizational Capacity (Skills, time) » Competing Projects » Heavy Reliance on IT » Tractors for Ethiopians – Need to learn how to use GIS – need the right tools	» Budget » Governance » Connectivity and IT Infrastructure » Environmental Issues (Cold, Rain, etc.) » Implement Wrong Hardware » Fatigue-Multiple Initiatives going on » Integration Problems » Buy-in » Staff Turn over	» Funding » End-user experience » Competing Projects » Lack of Buy-in from Decision Makers » Loss of Key Staff » Documentation	» Funding » Resources » Legal Threats - Liability » Data Quality Assurance/Quality Control » Infrastructure to access Data	» Funding » Knowledge » Staffing » Resources » Criticism (Public) » Inefficiency (GIS) » Political Commitment » Political Instability » Changing Priorities	» Lack of Funding » Departure of Key Staff » Buy-in-Competition for funds » Data quality (Esp. Public) » Not a Corporate Centralized System » Infrastructure - No Integration

FIGURE 5.8
SWOT threats.

- Ownership and commitment.
- Different perspectives on a complex problem versus a polarized take on issues or values.
- Individuals are willing to participate.
- The group has authority to make decisions.
- Creative solutions.
- Common goal or purpose.
- Trust.

Building consensus in any organization is essential. However, a key factor for success in any GIS strategic planning initiative is to achieve consensus not only within an organization but also with external entities and other interested parties. I have firsthand experience helping governments build consensus as it pertains to adopting and implementing GIS technology. I have listed a few examples to elucidate the process, as follows.

5.4.1 Example 1: San Luis Obispo County, California

San Luis Obispo County, California, chose to conduct a GIS needs assessment, develop a conceptual system design, and produce a strategic implementation plan. It was critically important to the county that the GIS action plan justify the expense of GIS implementation. Many activities supported consensus building. After project completion, 100% of the department directors agreed that the GIS was a vital technology. Each department *pushed in unison* to ensure that the plan was adopted. Consensus-building techniques were used to accomplish the overall goal of successful adoption and implementation. Figure 5.9 documents consensus-building techniques that were successfully used in San Luis Obispo.

5.4.2 Example 2: City of Carlsbad, California

The City of Carlsbad had a mature GIS but desired to take their GIS to an enterprise level. The city developed a two-pronged approach to GIS that included a specialized GIS group in information technology (IT) and public works. These two groups required guidance regarding the ways to ensure that individual goals were being achieved while still embracing organizational goals. The strategic planning initiative included numerous meetings to solicit feedback and build consensus. The strategic plan document allowed Carlsbad to identify the optimal growth for GIS while ensuring that each department's needs were met. Consensus-building techniques accomplished the overall goal of a successful adoption and implementation. Figure 5.10 documents the key consensus-building techniques that were successfully used in the City of Carlsbad.

SAN LUIS OBISPO, CALIFORNIA	
Methods and Techniques used by GTG to Build Consensus and Buy-In across the organization	The Real Benefit of Consensus Building or Collaborative Problem Solving in GIS Strategic Planning
1 Conducted **"Blue Sky Sessions"**	1 Improved quality of the GIS solution
2 Conducted **"Technical GIS Seminars"** for the entire organization	2 Better overall implementation
3 Conducted **"One-on-One"** interviews with each department	3 Allows innovative solutions
4 Used **"Case Studies"** and **"Real-World Demonstrations"** of applicable uses of GIS for specific departments	4 Minimizes the "impasse" factor
5 Used a **"Value Proposition and Return on Investment (ROI)"** approach or strategy to secure buy-in from decision makers.	5 Improves acceptance and willingness to participate
6 Used a **"Value Proposition and Return on Investment (ROI)"** approach or strategy to secure buy-in from decision makers.	6 Strengthens relationships between stakeholders
7 The **"4 Steps to enterprise-wide Implementation"**	7 A quantitative value and a return on investment – It saved money
8 Presentation of **"Governance Models"** used across the United States	8 Increases **"capacity building"** by removing obstacles to success
9 Open transparency of information	9 A clear outline, what needs to be decided

FIGURE 5.9
Consensus building case study—San Luis Obispo, California.

5.4.3 Example 3: Orange County, California

Each department in Orange County worked independently from each other. No central authority governed the GIS at the enterprise level, which led to stovepipes of information and an eclectic array of technologies. The strategic plan focused on creating an enterprise-wide vision for GIS, established a central GIS governance authority, standardized the technology and procedures, formalized expected RoIs, and attained buy-in from all departments. Consensus-building techniques were used to accomplish the overall goal of a successful adoption and implementation. Figure 5.11 documents the key consensus-building techniques that were successfully used in Orange County.

CITY OF CARLSBAD, CALIFORNIA			
Methods and Techniques used by GTG to Build Consensus and Buy-In across the organization		The Real Benefit of Consensus Building or Collaborative Problem Solving in GIS Strategic Planning	
1	Conducted **"Technical GIS Seminars"** for the entire organizations	1	Better Overall Implementation
2	Conducted **"One-on-One"** interviews with each department	2	Allows innovative solutions
3	Used "**Case Studies**" and "**Real World Demonstrations**" of applicable uses of GIS for specific departments	3	Minimizes the **"impasse" factor**
4	Conduct "**Blue Sky Sessions**"	4	Strengthens relationships between stakeholders
5	Presentation of "Governance Models" used across the United States	5	Increases **"capacity building"** by removing obstacles to success
6	**Live Software Demonstrations** with specific focus on how GIS can be used to solve the challenges of local government and local government utilities (Building and managing an Electric Utility System for example)	6	Better group relationships – through collaborative rather than competing stakeholders
7	**Open transparency of information**	7	A clear outline, what needs to be decided

FIGURE 5.10
Consensus building case study—City of Carlsbad, California.

5.4.4 Example 4: City of Dayton, Ohio

The City of Dayton had a mature, advanced GIS. However, the city did not have a GIS strategic plan to guide future activities. There was very little consensus across the enterprise. The strategic plan process introduced consensus building to ensure that all departments were working toward a common goal. On-site presentations and workshops brought the management team together. Through these presentations, the team gained a better, shared understanding of the individual and common GIS needs, the possible GIS uses, and a secured consensus on how best to move forward. Consensus-building techniques were used to accomplish the overall goal of a successful adoption and implementation. Figure 5.12 documents the key consensus-building techniques that were successfully used in the City of Dayton.

ORANGE COUNTY, CALIFORNIA	
Methods and Techniques used by GTG to Build Consensus and Buy-In across the organization	The Real Benefit of Consensus Building or Collaborative Problem Solving in GIS Strategic Planning
1 Conducted "**Blue Sky Sessions**"	1 Improves quality of the GIS Solution
2 Conducted "**Technical GIS Seminars**" for the entire organization	2 Better overall implementation
3 Conducted "**One-on-One**" interviews with each department	3 Allows innovative solutions
4 Used "**Case Studies**" and "**Real World Demonstrations**" of applicable uses of GIS for specific departments	4 Minimizes the "**impasse**" factor
5 Exceptional Presentation of the "**7 keys to GIS Success**"	5 Strengthens relationships between stakeholders
6 Presentation of "**Governance Models**" used across the United States	6 Increases "**capacity building**" by removing obstacles to success
7 **Open transparency of information**	7 A clear outline, what needs to be decided

FIGURE 5.11
Consensus building case study—Orange County, California.

CITY OF DAYTON, OHIO	
Methods and Techniques used by GTG to Build Consensus and Buy-In across the organization	The Real Benefit of Consensus Building or Collaborative Problem Solving in GIS Strategic Planning
1 Conducted "**Blue Sky Sessions**"	1 Improves quality of the GIS Solution
2 Conducted "**Technical GIS Seminars**" for the entire organizations	2 Better overall implementation
3 Conducted "**One-on-One**" interviews with each department	3 Allows innovative solutions
4 Used "**Case Studies**" and "**Real World Demonstrations**" of applicable uses of GIS for specific departments	4 Minimizes the "**impasse**" factor
5 Used a "**Value Proposition and Return on Investment (ROI)**" approach or strategy to secure buy-in from decision makers.	5 Improves acceptance and willingness to participate
6 Exceptional Presentation of the "**7 keys to GIS Success**"	6 Strengthens relationships between stakeholders
7 **Presentation of "Governance Models" used across the United States**	7 Increases "**capacity building**" by removing obstacles to success

FIGURE 5.12
Consensus building case study—City of Dayton, Ohio.

5.4.5 Example 5: Brant County, Ontario, Canada

A few departments in Brant County government utilized the GIS; however, no GIS staff nor a central GIS repository existed. A strategic plan allowed each department to gain an understanding of how the GIS could be utilized to assist with their unique needs. Consensus-building presentations and workshops were utilized throughout the county organization to ensure that all key staff members understood the county's GIS vision. Each department participated in achieving enterprise-wide success. Consensus-building techniques were used to accomplish the overall goal of a successful adoption and implementation. Figure 5.13 documents the key consensus-building techniques that were successfully used in Brant County, Ontario, Canada.

BRANT COUNTY, ONTARIO, CANADA	
Methods and Techniques used by GTG to Build Consensus and Buy-In across the organization	The Real Benefit of Consensus Building or Collaborative Problem Solving in GIS Strategic Planning
1 Conducted "Blue Sky Sessions"	1 Improves quality of the GIS Solution
2 Conducted "Technical GIS Seminars" for the entire organization	2 Better overall implementation
3 Conducted "One-on-One" interviews with each department	3 Allows innovative solutions
4 Used "Case Studies" and "Real World Demonstrations" of applicable uses of GIS for specific departments	4 Minimizes the "impasse" factor
5 Used a "Value Proposition and Return on Investment (ROI)" approach or strategy to secure buy-in from decision makers.	5 Improves acceptance and willingness to participate
6 Exceptional Presentation of the "7 keys to GIS Success"	6 Strengthens relationships between stakeholders
7 The "4 Steps to enterprise-wide Implementation"	7 A quantitative value and a return on investment – it saved money
8 Presentation of "Governance Models" used across the United States	8 Increases "capacity building" by removing obstacles to success
9 Live Software Demonstrations with specific focus on how GIS can be used to solve the challenges of local government and local government utilities (Building and managing an Electric Utility System)	9 Better group relationships – through collaborative rather than competing stakeholders

FIGURE 5.13
Consensus building case study—Brant County, Ontario, Canada.

GREENVILLE UTILITIES COMMISSION (GUC)		
Methods and Techniques used by GTG to Build Consensus and Buy-In across the organization		The Real Benefit of Consensus Building or Collaborative Problem Solving in GIS Strategic Planning
1	Conducted **"Blue Sky Sessions"**	1	Improves quality of the GIS Solution
2	Conducted **"Technical GIS Seminars"** for the entire organization	2	Better overall implementation
3	Conducted **"One-on-One"** interviews with each department	3	Allows innovative solutions
4	Used **"Case Studies"** and **"Real World Demonstrations"** of applicable uses of GIS for specific departments	4	Minimizes the **"impasse"** factor
5	Used a **"Value Proposition and Return on Investment (ROI)"** approach.	5	Improves acceptance and willingness to participate
6	The **"4 Steps to enterprise-wide Implementation"**	6	Strengthens relationships between stakeholders
7	**Open transparency of information**	7	A quantitative value and a return on investment – it saves money

FIGURE 5.14
Consensus building case study—GUC.

5.4.6 Example 6: Greenville Utilities Commission

Greenville Utilities Commission (GUC) is a large utilities organization in the eastern part of North Carolina. The organization needed to evaluate the accuracy of GIS field data collection and the needs of GUC's water and electric systems. It included an assessment of the geometric networks; the accuracy of all GIS data; data warehousing; and the formulation of a consensus-driven plan for improving the accuracy, maintenance, and effective utilization of the GIS data. The main components of the project included consensus building, Blue Sky sessions, small group interviews, presentations, questionnaires, and case study examples. Consensus-building techniques were used to accomplish the overall goal of successful adoption and implementation. Figure 5.14 documents the consensus-building techniques that were used during the development of a GIS strategic plan.

5.4.7 Example 7: City of Virginia Beach, Virginia

The City of Virginia Beach has central GIS staff in the IT department. The organization is large and very diverse, and most departments have their own GIS staff. One of the strategic plan project objectives was to gain consensus on the best technologies to manage the GIS in a way that met the needs of all departments. A number of consensus-building workshops focused on ensuring that each department understood the enterprise-wide GIS goals and that the results would yield solutions. Figure 5.15 documents the consensus-building techniques that were used during the development of a GIS strategic plan.

Figure 5.16 lists numerous methods and techniques that are used to promote consensus building within local government. Even though there is little empirical evidence to prove the value of consensus building, my experience in the business tells me that the building of consensus is a vital part of encouraging any organization to adopt and deploy GIS technology.

CITY OF VIRGINIA BEACH, VIRGINIA

Methods and Techniques used by GTG to Build Consensus and Buy-In across the organization	The Real Benefit of Consensus Building or Collaborative Problem Solving in GIS Strategic Planning
1 Conducted "Blue Sky Sessions"	1 Improves quality of the GIS Solution
2 Conducted "Technical GIS Seminars" for the entire organization	2 Better overall implementation
3 Conducted "One-on-One" interviews with each department	3 Allows innovative solutions
4 Used "Case Studies" and "Real World Demonstrations" of applicable uses of GIS for specific departments	4 Minimizes the "impasse" factor
5 Used a "Value Proposition and Return on Investment (ROI)" approach.	5 Improves acceptance and willingness to participate
6 Exceptional Presentation of the "7 keys to GIS Success"	6 Strengthens relationships between stakeholders
7 The "4 Steps to enterprise-wide Implementation"	7 A quantitative value and a return on investment – it saves money
8 Presentation of "Governance Models" used across the United States	8 Increases "capacity building" by removing obstacles to success
9 Live Software Demonstrations with focus on how GIS can be used to solve the challenges of local government and utilities	9 Better group relationships – through collaborative rather than competing stakeholders
10 Online Questionnaire used to solicit input from all stakeholders	10 Better decisions
11 Open transparency of information	11 A clear outline, what needs to be decided

FIGURE 5.15
Consensus building case study—City of Virginia Beach, Virginia.

METHODS & TECHNIQUES USED BY GTG TO BUILD CONSENSUS AND BUY-IN ACROSS THE ORGANIZATION	GTG CLIENTS						
	San Luis Obispo, CA	City of Carlsbad, CA	Orange County, CA	City of Dayton, OH	Brant County, Ontario, CA	Greenville Utilities Commission	City of Virginia Beach, VA
Conducting "Blue Sky Sessions"	●		●	●	●	●	●
Conducting "Technical GIS Seminars" for the entire organization	●	●	●	●	●	●	●
Conducting "One-on-One" interviews with each department	●	●	●	●	●	●	●
Using "Case Studies" and "Real World Demonstrations" of applicable uses of GIS for specific departments	●	●	●	●	●	●	●
Using a "Value Proposition and Return on Investment (ROI)" approach or strategy to secure buy-in from decision makers	●			●	●	●	●
Exceptional Presentation of the "7 Keys to GIS Success"	●	●	●	●	●	●	
The "4 Steps to Enterprise-Wide Implementation"	●					●	●
Presentation of "Governance Models" used across the United States	●	●	●	●			●
Live Software Demonstrations with specific focus on how GIS can be used to solve the challenges of local government and local government utilities (building and managing an Electric Utility System for example)						●	●
Online Questionnaire used to solicit input from all stakeholders							●
Open transparency of information		●	●			●	

FIGURE 5.16
Methods and techniques for consensus building.

THE REAL BENEFITS OF CONSENSUS BUILDING OR COLLABORATIVE PROBLEM SOLVING IN GIS STRATEGIC PLANNING	GTG CLIENTS						
	San Luis Obispo, CA	City of Carlsbad, CA	Orange County, CA	City of Dayton, OH	Brant County, Ontario, CA	Greenville Utilities Commission	City of Virginia Beach, VA
Improve Quality of the GIS Solution	●		●	●	●	●	●
Better Overall Implementation	●	●	●	●	●	●	●
Allow Innovative Solutions		●	●	●	●	●	●
Minimized the "Impasse" Factor	●	●	●	●	●	●	●
Improved Acceptance and Willingness to participate	●			●	●	●	●
Strengthens Relationships between stakeholders	●	●	●	●	●	●	●
A Quantitative Value and a Return on Investment - It Saved Money	●				●	●	●
Increased Capacity Building by removing obstacles to success	●	●	●	●	●		●
Better Group Relationships - through collaborative rather than competing stakeholders					●	●	●
Better Decisions							●
A Clear Outline, what needs to be decided		●	●			●	●

FIGURE 5.17
Benefits of consensus building.

Figure 5.17 presents the benefits of consensus building. Consensus building could also be called collaborative problem-solving. It can resolve conflicts within an organization. It will allow a GIS stakeholder community to work together to develop a mutually acceptable GIS initiative. Consensus building is a critical component of your success.

5.5 Step Four: Align the GIS with the Vision of the Organizations

One of your main objectives is to organize your GIS initiative so that it aligns with your organization's overall mission, vision, goals, and objectives. This is probably the most overlooked and critically important part of any GIS strategic planning initiative. Figure 5.18 is an example of a city's corporate strategic framework. Using only a page, this city highlights the overall vision of the organization. This is an excellent example of developing a vision for the organization. It is essential that you review and embrace this before developing any goals for your enterprise GIS.

Most local government municipalities recognize that strategic planning is good practice and can transform the organization. An organization's overall (non-GIS) vision statements tend to talk about providing high-quality, innovative, and cost-effective services that support the lives of citizens and the vitality of neighborhoods. Some talk about their mission, guiding principles, and objectives. Most include the following language:

- Quality service
- Innovation
- Professionalism
- Teamwork
- Integrity
- Resourcefulness
- Continually improve the level of assistance to the public
- Accuracy
- Continuous improvement to assist the public
- Make departmental procedures *user friendly*

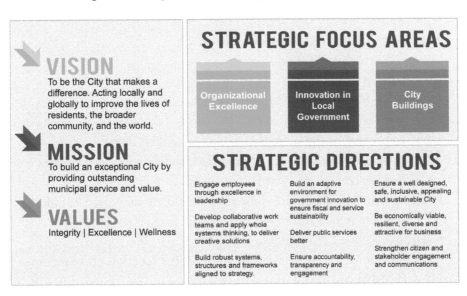

FIGURE 5.18
Example of a vision, goals, and objectives.

Your job as a GIS coordinator is to frame the GIS vision, goals, and objectives so that they use this type of language, and they are synchronized or *aligned* with the organization's overall mission, goals, and objectives. Let us have a look at one organization's GIS alignment strategy. This is a good example of how a local government GIS coordinator embraces the idea of aligning their GIS goals with the overall vision of the organization. There are six organizational goals, including the following:

1. Continually improve public safety
2. Sound and stable utilities
3. Promote economic development
4. A great downtown
5. Increased civic engagement
6. Promote fiscal soundness

Let us see how this example city has tried to align the GIS goals with the overall goals of the organization.

5.5.1 Organization's Overall Goal #1: Continually Improve Public Safety

The GIS division's activity to support this organizational goal is as follows:

- Deploy GIS crime analysis software.
  - Reduce crime.
- Deploy GIS with computer-aided dispatch.
  - Incident visualization.
  - Closest unit dispatch.
- Deploy GIS Internet public access and public engagement solutions.
  - Improve public notification.
  - Improved information dissemination to the public.
  - *Alert-360*—real-time situational awareness for first responders.
  - Engage community watch.
- Deploy GIS-centric fire incident visualization and analysis solution.
  - Identify trends and patterns to optimizing resources.
- Deploy Mobile GIS solutions.
  - Mobile GIS adds immediate geographic location and efficient routing.

- Utilize GIS for real-time *roll call* mapping.
  - Enable all police officers to view incident data on a regular basis.
- Deploy GIS-centric executive dashboards.
  - Identify patterns and trends.
  - Monitor crime in real time.
- Deploy GIS tools in the emergency operations center (EOC).
- Create and maintain critical GIS data.

5.5.2 Organization's Overall Goal #2: Sound and Stable Utilities

The GIS division's activities to support this organizational goal are as follows:

- Deploy GIS-centric asset management tools.
  - Tracking, managing, locating, and navigating to critical infrastructure.
- Integrate GIS with work order solutions.
  - Optimum use of field resources.
- Integrate GIS with Capital Improvement Program (CIP).
  - Improve awareness of competing projects and tasks.
  - Monitor CIP projects and inform the public.
- Infrastructure auditing and depreciation with GIS.
  - Improve Government Accounting Standards Board Statement 34 (GASB34) accounting and depreciation.
- Utilize GIS as a decision support tool.
  - Predict infrastructure failure.
- Integrate GIS with billing software.
  - Visualize consumption patterns and efficiency.
- Deploy GIS-centric crowdsourcing applications.
  - Enable citizens to repost problems.
  - Improve public awareness involvement.
- Deploy GIS-centric mobile field access.
  - Reduce duplication of effort.
  - Automate a traditionally paper-based process.
  - Route optimization.
- Well-maintained city streets, roads, parks, and recreation centers help protect property values and maintain the quality of life.

5.5.3 Organization's Overall Goal #3: Promote Economic Development

The GIS division's activity to support this organizational goal deploys the following:

- Internet Economic Web Portal.
    - Improve public access to local data.
    - Offer automated geographic site selection.
    - Promote local economy.
    - Encourage retention.
- Integrate GIS with economic opportunity and available property database.
- Integrate extensive demographic data into GIS.
    - Enable prospective business with geographic tools.
- Deploy simple interactive story maps.
    - Promote available land and office space.

5.5.4 Organization's Overall Goal #4: A Great Downtown

The GIS division's activity to support this organizational goal is as follows:

- Use GIS to support mapping and decision support associated with the downtown plan.
    - GIS mapping to illustrate street improvement and housing projects.
- Improve public awareness through online interactive maps.
    - Promote the downtown amenities and businesses.
    - Storyboard large events.
    - Cultural arts and a public art walking tours.

5.5.5 Organization's Overall Goal #5: Increased Civic Engagement

The GIS division's activity to support this organizational goal is as follows:

- Deploy 311 GIS crowdsourcing solutions.
    - Improve public involvement.
- Plan, design, and deploy story maps.
    - Large events.
    - History of the organization.
    - Economic opportunities.

- Deploy GIS Internet solutions.
  - Promote government services.
  - Show crime and crime patterns.
  - Promote economic development.
- Use social media and GIS.
  - Mapping ideas and complaints of citizens.

5.5.6 Organization's Overall Goal #6: Promote Fiscal Soundness

The GIS division's activity to support this organizational goal is as follows:

- Utilize GIS to save money and avoid costs.
- Use GIS to save time.
- Deploy GIS to increase productivity and organizational performance.
- Utilize GIS technology to improve efficiency.
- GIS technology will improve data accuracy and reliability.
- GIS is useful in making better and more informed decisions.
- GIS for saving lives and mitigating risks.
- GIS for automating workflow procedures.
- GIS solutions can improve information processing.
- GIS will help comply with state and federal mandates.
- GIS will protect the community and save money.
- GIS will assist in improving communication, coordination, and collaboration.
- GIS will provide data to regulators, developers, and other interested parties.
- GIS will help respond more quickly to citizen requests.
- Improve citizen access to government using GIS.
- Effective GIS management of assets and resources.
- GIS helps with good environmental stewardship and well-being.
- GIS will promote economic vitality.

Goal alignment is a strategic initiative and a key component of a GIS strategic plan. It helps the GIS staff work toward the local government's overarching goals. Proper alignment will ensure that the GIS hits the mark and allows talented workers to be more effective. It is important for GIS staff to know why GIS alignment is important, why real-time feedback and tracking is a good practice, and why it helps with communication.

5.6 Step Five: Create a GIS Vision Statement, Goals, and Objectives

We now understand the existing GIS situation within the organization. We have conducted Blue Sky GIS sessions with stakeholders, and we have conducted a GIS SWOT analysis. We have responsibly built consensus and buy-in for an enterprise GIS solution from all stakeholders, and we have attempted to align our GIS initiative with the overall vision of the organization. This leaves us with two tasks. We need to develop a GIS vision statement, goals, and objectives and develop performance measures.

5.6.1 GIS Vision Statement

The objective is to review all the information that is gathered during the previous steps and develop a vision statement that goes something like this: *develop and maintain an enterprise, scalable, and sustainable GIS that promotes effective and innovative use of geospatial technology, supported by good GIS governance and coordination; standards; and ongoing training, education, and knowledge transfer.* Figure 5.19 illustrates how diverse local government vision statements can be.

LOCAL GOVERNMENT GIS VISION STATEMENTS

To become the premier provider of spatial information and GIS services in the region.

The mission of County GIS is to empower County's officials, employees and partners, and support data-driven decision making throughout the agency by facilitating the development, use and interpretation of high quality geospatial data.

The GIS Vision of the Town is to establish an organizational structure that allows Town staff to effectively maintain and use geo-spatial data, and also allows the citizens efficient access to pertinent data. This organization will be consistent with Council goals and Town policy so that the data is accurate, reliable, and consistent.

GIS will be an easy-to-use decision-making tool available on all County computers that enables access to data and information, promotes innovative solutions, and improves customer service to residents.

The vision and mission of the City GIS staff is to support the activities of the City and its citizens by providing and maintaining accurate, current and complete geospatial data. This support will be provided through leveraging the knowledge contained in this information by using a set of procedures and techniques collectively referred to as a Geographic Information System. Using the Geographic Information System (GIS), the staff will enable the managers and citizens to make decisions impacting the future of the City in an informed and logical manner.

Provide a robust and high quality geographic information system that empowers users to efficiently access, manage, maintain, and share accurate, reliable, and consistent geographic data; and to easily and quickly analyze and obtain information in various formats on demand.

FIGURE 5.19
Example of local government vision statements.

5.6.2 GIS Goals

The next step in the process is to develop GIS goals. These goals need to focus on the main components that are identified in the formula for success outlined in Chapter 3, including (a) governance, (b) data and databases, (c) procedures and workflow, (d) GIS software, (e) training, (f) education and knowledge transfer, and (g) GIS infrastructure. Figure 5.20 illustrates my approach to goal setting that is tailored around the key components of an enterprise GIS.

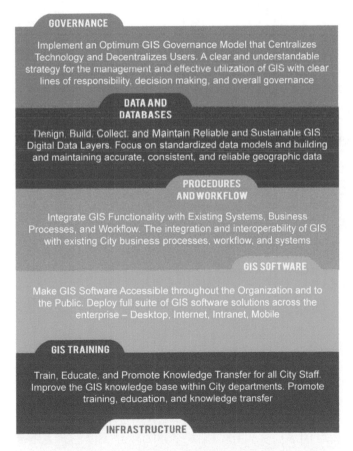

GOVERNANCE

Implement an Optimum GIS Governance Model that Centralizes Technology and Decentralizes Users. A clear and understandable strategy for the management and effective utilization of GIS with clear lines of responsibility, decision making, and overall governance

DATA AND DATABASES

Design, Build, Collect, and Maintain Reliable and Sustainable GIS Digital Data Layers. Focus on standardized data models and building and maintaining accurate, consistent, and reliable geographic data

PROCEDURES AND WORKFLOW

Integrate GIS Functionality with Existing Systems, Business Processes, and Workflow. The integration and interoperability of GIS with existing City business processes, workflow, and systems

GIS SOFTWARE

Make GIS Software Accessible throughout the Organization and to the Public. Deploy full suite of GIS software solutions across the enterprise – Desktop, Internet, Intranet, Mobile

GIS TRAINING

Train, Educate, and Promote Knowledge Transfer for all City Staff. Improve the GIS knowledge base within City departments. Promote training, education, and knowledge transfer

INFRASTRUCTURE

Build and Maintain an Enterprise IT Infrastructure to support an Enterprise GIS. Architecture and enterprise initiative that will sustain growth and change

FIGURE 5.20
GIS goals.

GOVERNANCE

VISION: Should implement an Optimum GIS Governance Model that Centralizes Technology and Decentralizes Users

GOAL: Requires an understandable strategy for the management and effective utilization of GIS with clear lines of responsibility, decision making, and overall governance

TASKS	KEY PERFORMANCE MEASURES
Task 1: GIS Authority	» ACC should create a GIS Coordinator position within the CIS Department to oversee all GIS functions within the County. There is a distinct need for a GIS Coordinator within the CIS department that supports all departments equally, and has the authority to direct GIS development across the enterprise.
Task 2: Accessibility	» Improve accessibility to GIS software and GIS data to all county employees and other interested parties.
Task 3: Strategic Plan	» Develop a Three-Year GIS Strategic Plan that details every task required to deploy a true enterprise solution that takes advantage of the latest technology and architecture.
Task 4: Standard Operating Procedures	» Establish a set of standards and procedures for the development and maintenance of geospatial data including Office-to-Field – Field-to-Office, GPS Quality Standards, Versioning, CAD Standards, Digital Submission Standards, Cartographic Standards, Metadata Standards, Standard Naming Conventions, and GIS Business Integration. Replicate the County's existing SOP's approach for addressing and Parcel maintenance
Task 5: Grants and Funding	» The GIS Coordinator must pursue grants for GIS software, data, training, and staff.
Task 6: Coordinated Enterprise	» Develop and coordinate an enterprise GIS training plan administered by the GIS Coordinator.
Task 7: Create a GIS Steering Committee	» Create a GIS Steering Committee comprised of directors from each department. Develop a strategy for the Steering committees goals and objectives.
Task 8: Develop a GIS user Group	» The proposed GIS Steering Committee should select who participates in the GIS User Group. Develop a strategy for the GIS User Groups goals and objectives.
Task 9: Formalize the Governance Model	» Adopt a new governance model that places the GIS coordinator in CIS. The existing decentralized model prevents a true enterprise use of GIS.
Task 10: Performance Measures	» Establish enterprise and departmental GIS performance measures and Return on Investment (ROI)
Task 11: GIS Newsletter	» Create a quarterly City-County-wide newsletter, blog, and tweet, to promote GIS and showcase successes
Task 12: Update to the Plan	» Update the GIS Strategic Plan on an annual basis using an online questionnaire and departmental interviews
Task 13: GIS Coordination Tasks	» Develop a document that details each departments GIS needs. The GIS coordinator and support staff will perform these tasks on a weekly, monthly, and annual basis. This will include technical support and project management.

DATA AND DATABASES

VISION: Should Design, Build, Update, Collect, and Maintain Reliable and Sustainable GIS Digital Data Layers

GOAL: Should use Esri's Local Government Information Model (LGIM) as the standardized data model for future growth. A modified or enhanced LGIM should be used to build and maintain accurate, consistent, and reliable geographic data.

TASKS	KEY PERFORMANCE MEASURES
Task 1: Digital Data Assessment	» Perform a comprehensive assessment of the quality, quantity, and completeness of all digital data layers.
Task 2: Enterprise Database Design	» Establish and implement Esri's LGIM – with opportunities for customization where needed
Task 3: Data Creation	» Develop a plan to collect, update, and convert all required departmental digital data layers over an agreed upon period of time including the list of required data layers detailed in the GIS Assessment – including but not limited to (1) signs, (2) Knox box locations, (3) hydrants, (4) bike lanes and more.

FIGURE 5.21

GIS objectives and performance measures.

(Continued)

DATA AND DATABASES (Continued)

VISION: Should Design, Build, Update, Collect, and Maintain Reliable and Sustainable GIS Digital Data Layers

GOAL: Should use Esri's Local Government Information Model (LGIM) as the standardized data model for future growth. A modified or enhanced LGIM should be used to build and maintain accurate, consistent, and reliable geographic data.

TASKS		KEY PERFORMANCE MEASURES
Task 4: Improve Access to Critical Foundation Data Layers	»	Develop a multi-tiered approach allowing all departments to view and access base map layers including parcels, address points, street centerlines, and aerial photography in real-time. ACC should continue using qPublic as the browser of choice, but supplement this with ArcGIS Online tools.
Task 5: Master Data List	»	Update ACC's existing master digital data list and maintain on a regular basis.
Task 6: Central Repository	»	Continue to utilize Esri's ArcSDE environment to house all City- County GIS data. This central repository is located within the CIS Department
Task 7: Metadata	»	Establish and enforce standard operating procedures (SOP) for developing metadata standards
Task 8: Custodianship	»	Clearly define data custodianship roles within the enterprise governance model. This includes coordination between all departments.
Task 9: Mobile Solutions for Viewing and Maintaining Data	»	Plan, design, and deploy Esri's ArcGIS Online as the mobile software solution for ACC. This should be supplemented by a continued effort to identify new, advanced, convenient, and easy-to-use mobile GIS and GPS field tools to collect, update, and maintain ACC GIS data repository. A uniform approach to using ArcGIS Online would benefit ACC

PROCEDURES AND WORKFLOW

VISION: Promote the interoperability of GIS with ACC's existing business systems

GOAL: Integrate GIS Functionality with Existing Database Systems, Business Processes, and Workflow

TASKS		KEY PERFORMANCE MEASURES
Task 1: Integration	»	Integrate GIS with ACC's existing business systems including, but not limited to: • qPublic (missing GIS data layers) • FireHouse • WinGAP • Pictometry • Intergraph • Laserfiche • Energov • Microsoft Excel • AutoCAD • Microsoft Access
Task 2: Opportunities and Gaps	»	Integrate GIS into departmental business workflows – • Public Web Portal (Leisure Services, police, Tax Assessor, Transportation, Public Works and Planning • City-County-wide Intranet Solution - qPublic, ArcGIS Online Applications (All Departments) • CAD Standards (Public Utilities, Transportation and Public Works) • Desktop Analysis and Modeling Extensions (Public Utilities, Transportation and Public Works)
Task 3: Mobile Solutions	»	Integrate ArcGIS Online Mobile into departmental workflow procedures. ACC will be required to Plan, design, and deploy Esri's ArcGIS Online as the mobile software solution for the enterprise. This should be supplemented by a continued effort to identify new, advanced, convenient, and easy-to-use mobile GIS and GPS field tools to collect, update, and maintain the City-County's GIS data repository.

(Continued)

FIGURE 5.21 (CONTINUED)
GIS objectives and performance measures.

GIS SOFTWARE

VISION: Make GIS Software Accessible throughout the Organization, and to the Public and other interested parties

GOAL: Deploy a full suite of Esri GIS software solutions across the enterprise – Desktop, Internet, Intranet, and Mobile.

TASKS	KEY PERFORMANCE MEASURES
Task 1: Enterprise License Agreement	» Upgrade ACC's existing license agreement with Esri to an Enterprise License Agreement. Effectively utilize Esri's software by responsibly deploying the right tools to the right people.
Task 2: Intranet Solution	» Continue to utilize qPublic and the Flex Viewers to provide staff with live views of GIS data. ACC should evaluate moving to Esri's HTML5 viewer during Q1 of 2015.
Task 3: ArcGIS Online (AGOL) Software Initiative	» Plan, design, and deploy AGOL, including the setup, configuration, and effective use of the tools and applications available.
Task 4: Story Maps	» Develop a sequence of ACC wide story maps using a five step process including 1. Storyboarding, 2. Data Gathering, 3. Design, 4. Build and Refine, and 5. Publish and Maintain. Possible Story Boards include: 1. History of Athens-Clarke County 2. ACC Leisure Services 3. Downtown Athens-Clarke County 4. Athens-Clarke County Events
Task 5: Internet Public Access Portal	» Improve public access to online ACC services. AGOL should be considered as a viable solution for the Public. The following departments will have a role to play in the public access portal initiative Planning, Transportation and Public Works, Leisure Services, and Tax Assessor.
Task 6: Crowdsourcing	» Engage and solicit input from citizens by promoting crowdsourcing applications. Utilize a reliable database to house information gathered from the crowdsourcing application.
Task 7: ACC Commissioners	» Use GIS as a tool to provide timely and accurate data to ACC Commissioners.
Task 8: Modeling Extensions	» ACC should take advantage of Esri's modeling extensions for the desktop, Public Utilities, Transportation and Public Works, and other departments, can take advantage of 3D Modeling, Spatial Analyst, Network Analyst, and City Engine.
Task 9: Mobile Software Solutions	» Plan, design, and deploy Esri's ArcGIS Online as the » mobile software solution for ACC. This should be supplemented by a continued effort to identify new, advanced, convenient, and easy-to-use mobile GIS and GPS field tools to collect, update, and maintain the City-County's GIS data repository.

GIS TRAINING

VISION: Train, Educate, and Promote Knowledge Transfer for all ACC Staff

GOAL: Improve the GIS knowledge base within ACC departments. Develop a training, education, and knowledge transfer plan. Encourage the effective utilization of GIS technology.

TASKS	KEY PERFORMANCE MEASURES
Task 1: Governance Model	» Implement a centralized hybrid governance model that promotes ongoing training and education.
Task 2: Software Training	» Provide software GIS training and educational opportunities to all ACC staff on a regular basis. Utilize Esri's Online Education and Training services through the purchase of the ELA. Provide formal classroom training for identified departmental staff – including Desktop, Intranet, Internet, Mobile, GPS, ArcGIS Online and Story Maps, and Extensions.

(Continued)

FIGURE 5.21 (CONTINUED)
GIS objectives and performance measures.

GIS TRAINING (Continued)

VISION: Train, Educate, and Promote Knowledge Transfer for all ACC Staff

GOAL: Improve the GIS knowledge base within ACC departments. Develop a training, education, and knowledge transfer plan. Encourage the effective utilization of GIS technology.

TASKS	KEY PERFORMANCE MEASURES
Task 3: Knowledge Transfer	≫ Establish a GIS user group network within the organization to help facilitate growth. Establish quarterly GIS meetings.
Task 4: Formal On-going Training Plan	≫ Implement a formal sustainable GIS Training Plan. Task #5: MOBILE TRAINING – As part of the formal training plan, develop a strategy for the effective use and training of mobile field devices.
Task 5: Conferences	≫ Attend workshops and pre-conference seminars at the Esri International Users Conference and regional Esri Conferences, such as the SERUG.
Task 6: Online Seminars and Workshops	≫ Use all available online training, education, and knowledge transfer workshops.
Task 7: Brown Bag Lunches	≫ The GIS Coordinator will offer seminars and workshops tailored to specific departmental applications of GIS.
Task 8: Departmental Specific Training	≫ Promote departmental specific GIS training.
	Promote departmental specific GIS training. Encourage and promote targeted GIS training for ACC departments, including : General Management Workshop | Public Safety GIS Workshop | Utilities Workshop | The ROI of GIS in City-County Government
Task 9: Departmental Specific Training	≫

- **INTRODUCTION TO GIS** - An Introduction to Geographic Information Systems (GIS) and Local Government - What is GIS?
- **OUR VISION FOR GIS** - Vision and Goals and Objectives for our Enterprise and Sustainable GIS- What do we want to accomplish?
- **GIS CHALLENGES** - The Challenges, Barriers and Pitfalls of GIS in Local government - What problems may we encounter and how can we fix them?
- **GIS MANAGEMENT** - GIS Management and Governance – How to effectively manage and maintain our GIS
- **DATA VS INFORMATION** - Our Digital World: Turning Local, State, and Federal Data into Meaningful Information – How do we effectively use the abundance of digital data?
- **GIS SOFTWARE APPLICATIONS** - Local Government Departmental GIS Applications and Opportunities - How are local government departments using GIS technology?
- **GIS ARCHITECTURE** - Enterprise GIS Architecture: Building an enterprise solution that includes hardware, networking and mobile applications.
- **GIS: THE RETURN ON INVESTMENT (ROI)** - What's our Business Case, Return on Investment (ROI) and Value Proposition of an Enterprise GIS? – Quantifying the value and learning how to sell GIS
- **ESRI'S GIS SUITE OF SOLUTIONS** - The Esri GIS Suite of Software Solutions - Understanding the different GIS solutions from sophisticated desktop solutions to mobile tools, and intranet applications.
- **THE FUTURE OF GIS IN LOCAL GOVERNMENT** - The Future of GIS Technology and Geospatial Science- What can we expect in the next 10 years?
- **GOOD GIS BUSINESS PRACTICES** - GIS: Best Business Practices (BBP) from around the World – Learning from other organizations.
- **THE CITY OF ROSWELL'S GIS IMPLEMENTATION** - The City of Roswell's Enterprise GIS – How successful is the City of Roswell in building an enterprise GIS?

FIGURE 5.21 (CONTINUED)
GIS objectives and performance measures.

LOCAL GOVERNMENT (COUNTY)

Develop an enterprise, scalable, and sustainable GIS that promotes effective and innovative use of geospatial technology, supported by good GIS governance and coordination, standards, and on-going training and education.

COUNTY GOVERNMENT GOALS

Goal A	Goal B	Goal C	Goal D	Goal E	Goal F
GOVERNANCE Should implement an Optimum GIS Governance Model that Centralizes Technology and Decentralizes Users	**DATA AND DATABASES** Should Design, Build, Update, Collect, and Maintain Reliable and Sustainable GIS Digital Data Layers	**PROCEDURES AND WORKFLOW** Promote the interoperability of GIS with existing business systems	**GIS SOFTWARE** Make GIS Software accessible throughout the Organization, and to the Public and other interested parties	**GIS TRAINING** Train, Educate, and Promote Knowledge Transfer for all Staff	**INFRASTRUCTURE** Should Continue to Utilize the IT Infrastructure to support an Enterprise, Scalable and Sustainable GIS
Requires an understandable strategy for the management and effective utilization of the GIS with clear lines of responsibility, decision making, and overall governance.	Should use Esri's Local Government Information Model (LGIM) as the standardized data model for future growth. A modified or enhanced LGIM should be used to build and maintain accurate, consistent, and reliable geographic data.	Integrate GIS Functionality with Existing Database Systems, Business Processes, and Workflow.	Deploy a full suite of Esri GIS software solutions across the enterprise – Desktop, Internet, Intranet, and Mobile.	Improve the GIS knowledge base within departments. Develop a training, education, and knowledge transfer plan. Encourage the effective utilization of GIS technology.	Continually evaluate Architecture initiative so it will sustain enterprise growth and change.

COLUMBUS CONSOLIDATED GOVERNMENT OBJECTIVES

FIGURE 5.22
GIS vision, goals, and objectives.

5.6.3 GIS Objectives

GIS objectives or tasks are the third tier of our strategic goal setting. GIS objectives are the details that will set the scene for performance measures. Establishing GIS objectives allows us to create effective service level agreements between the GIS group and each department; memorandums of understanding between departments; and a scope of services, schedule, and priorities and establish GIS performance measures that will allow you to monitor and quantify success. Figure 5.21 presents the GIS objectives and the performance measures.

5.7 Step Six: Develop Performance Measures, Outcomes, and Metrics

Elected officials and decision makers require evidence that progress is being made. Performance measures essentially measure the performance of the GIS coordinator, the GIS user group, the GIS steering committee, and the entire organization. Figure 5.21 illustrates how the GIS vision statement, GIS goals, GIS objectives or tasks, and performance measures can be combined using the components that are detailed in the formula for success. The performance measures are a tool to quantify the tasks that are outlined in the strategic plan.

The end product of these sequential steps is illustrated in Figure 5.22. This approach allows all stakeholders to see the vision, goals, and objectives on one page.

6

Governance

Lead me, follow me, or get out of my way.

General George Patton

6.1 Introduction

At this phase in the discussion, the question we have to ask ourselves is unrelated to the data-management capabilities of GIS technology or its potential benefit. We have been through all of that. Rather, the question pressing down on us now is, how do small towns, midsized cities, or incredibly large counties manage this technology called GIS? How do local government organizations juggle all of the complex strategic, technical, tactical, logistical, and political issues that accompany such powerful technology?

We would be well advised to remember what Baghai, Coley, and White stated in *The Alchemy of Growth*,

> We live in an era rich in opportunities. Our experience suggests that growth prospects are limited more often by management failings than by economic realities. The question for underperforming companies is thus not whether growth is possible, but whether they are prepared to take on the growth challenge.

6.2 But What Exactly Is GIS Governance?

GIS governance refers to all of the processes and actions that are required to manage the planning, design, implementation, and ongoing maintenance of GIS technology. Governance is integrated throughout all of the components of GIS. GIS governance is the management of an integrated solution that serves an entire organization by offering levels of geospatial functionality, uniform standards, good management, reliable digital data and databases, workflow procedures, training education and knowledge transfer, and a backbone for architecture and infrastructure.

GIS governance is a social and political process, and the aggregated experiences of local government organizations prove that achieving an operational GIS does not guarantee its use. The correct implementation of an appropriate governance model can give rise to positive and beneficial characteristics in an organization. Conversely, selecting and/or implementing a poorly suited governance model that does not follow the implementation principles can have negative consequences. The ultimate success of an enterprise-wide GIS will depend on the ability to govern and manage the GIS in an evolving multidepartmental environment.

6.3 New Management Challenges Introduced by GIS Implementation

There are no two ways about it GIS technology, by its very enterprise and sophistication, brings an entirely new set of management challenges to an organization. These changes can disrupt, for better and worse, the processes that an organization relies on for effective and democratic decision making, consensus building, organization-wide resource planning, project and process management, project prioritization, data maintenance, and accountability, not to mention, of course, the financial challenges of deploying GIS technology or the added layer of complexity that the human element inevitably brings. The personality types and agendas of local government officials, along with the relationship between administrators and the users of GIS technology, are as nuanced and important to understand as any facet of the technology.

6.4 Government without GIS

Before we get into the meat of our discussion about good GIS governance, let us take a look at what would happen if an organization does not deploy GIS technology. After all, if we consider an organizational landscape without the GIS, we can isolate the components of GIS implementation that will present new management challenges. In the modern world, an organization that does not use a GIS would have some, if not all, of the following characteristics that are detailed in Figure 6.1. The list in Figure 6.1 illustrates what a streamlined twenty-first-century government organization *does not look like*. This kind of unwieldy and ineffective governance is an outcome of not investing in a GIS.

GOVERNMENT WITHOUT GIS

☑ A predominantly manual decision support process.
    (Remember we did manage to survive without cell phones,
    texting, and email, but at what cost?
☑ Duplication of effort
☑ Unmet technology expectations of staff
☑ Wasted time, energy, and resources
☑ Inefficient workflow processes
☑ Missinformed or non-informed public
☑ A complicated and unorganized compilation of map data
☑ A failure to effectively turn geospatial data into meaningful
    and effective information
☑ Less than effective communication and collaboration
☑ Neither a streamlined or modernized system
☑ Lack of open and transparent government
☑ Poor citizen interaction and feedback
☑ Poor management of resources and assets
☑ No map data standards
☑ No predictive analytical capabilities

FIGURE 6.1
Government without GIS.

6.5 Misguided GIS Governance

What about organizations that have invested in GIS technology? What about misguided governance and the damage that this has done on a local government organization? Arguably, we could say that deploying and managing a true enterprise GIS is not an easy undertaking. Because of its complexity, the outcomes can often be less than successful. Figure 6.2 enumerates the possible outcomes of misguided GIS governance.

OUTCOMES OF MISGUIDED GOVERNANCE

☑ Assumption based decision making
☑ Empire building
☑ Misinformed public
☑ Data and process duplication
☑ Variations in priorities
☑ Constant internal competition over funding projects and
    resources
☑ Information hording or missing information
☑ Inability to locate critical or timely information
☑ Insensitivity to users' needs
☑ Insufficient prevention and response
☑ Inefficient decision making
☑ Poor training and education
☑ Poorly maintained, misplaced and stale information
☑ Everyone going their own way

FIGURE 6.2
Outcomes of misguided GIS governance.

6.6 Why Is It So Challenging to Deploy an Optimum GIS Governance Model That Meets a Set of Basic Criteria?

The outcome of an ill-defined, haphazard governance model leads to poor GIS service delivery and an unsustainable management strategy. Though the GIS is a technology that rapidly becomes taken for granted among government professionals, I must concede that managing a GIS is a far more complex problem than managing any other system. GIS technology involves virtually every department in an organization and engages a massive number of people across multiple tiers of users. Local government organizations, whether town, city, or county, often find it extremely complex and problematic to deploy a sustainable GIS governance model that meets a specific set of performance criteria. This begs the question, why is GIS much harder to manage and maintain than other traditional technologies? Let us take a look at the top 12 reasons that local government organizations have historically struggled with good GIS governance. Figure 6.3 illustrates 12 reasons why GIS is complicated.

1. *The number and type of GIS users in an organization make it far more complex to manage than any other comparable technology.* More often than not, the GIS is underutilized throughout an organization, and a sheer number and diversity of user types can present complex managerial problems.

2. *The number and complexity of applications can make GIS overwhelming.* We all know that GIS technology is multifaceted. From a simple public browser to sophisticated predictive modeling algorithms, the technological tasks required to manage desktop, server, Internet, Intranet, mobile, and online cloud solutions present yet another challenge to GIS governance.

3. *Organization-wide agendas, priorities, and constraints.* We should not be naive about government operations. The different departmental and organization-wide agendas and priorities play a significant role in the deployment of every governmental plan. GIS is no different, and budget constraints will, more often than not, determine the level of GIS success.

4. *The surprising "job description factor" that GIS comes bundled with can be puzzling.* One of the barriers to good management and the maintenance of GIS in local government is the painfully slow adoption of GIS job descriptions, or the acceptance of redefining job descriptions to embrace geospatial technology. A large section of government employees effectively uses GIS technology for a substantial amount of their workday without GIS being in their formalized job descriptions, and thus the vocabulary surrounding a job description is highly important. One would not want to scare government officials with unnecessary semantics.

12 REASONS WHY GIS GOVERNANCE IS COMPLICATED					
1 Types of Users • Number and types of users makes GIS more complex to manage than other technologies • Career- Analytical- Browsers- Casual	**2 Applications and Software** • Number and complexity of applications make GIS overwhelming • Departmental Business Needs • Application Development • Open Source	**3 Priorities and Alignment** • Departmental & organization-wide agendas factor into every plan • Budget constraints • Vision, Goals, and Objectives • Alignment	**4 Job Descriptions** • Redefining job descriptions to embrace geospatial technology • GIS being in formalized job descriptions is important	**5 Education, Training, and Knowledge Transfer** • GIS training plan should be supplemented with employee education and activities • Seminars, conferences	**6 Citizen Engagement** • Open and transparent government • Crowdsourcing initiatives • Citizen engagement • Government accountability
7 People and Personality • Local government employs a wide variety of professionals • Technology perspective- management perspective- cultural perspective	**8 GIS Funding** • Budget and funding are major factors • Enterprise, operational, and capital funds • Funding for GIS requires constant monitoring • Budget- Grants- Payback Models	**9 Workflow Complexity** • Standard Operating Procedures (SOP's) help educate staff • Detailed understanding of business tasks • Integration and interoperability	**10 Mobility, Technology, & Architecture** • Deploying the correct hardware, networking, software, and operating systems • "Office to field" & "field to office" capabilities	**11 Historical Adoption** • Non-linear history of GIS deployment • Decentralized approach • Organic growth can lead to disregard for enterprise needs • Centralized model is created	**12 Data- Open Data & Open Government** • State- Federal- Local- Private- Accuracy, inaccuracy, and complete • Publishing and Sharing

FIGURE 6.3

Difficulties of GIS—12 reasons why GIS governance is complicated.

5. *Education, training, and knowledge transfer.* A GIS training plan for your organization needs to be supplemented with thorough employee education and activities that are geared toward the transmission of knowledge. Mobilizing and training staff across different departments are never easy.

6. *The public involvement and engagement factor.* The unparalleled ability of GIS software solutions to support open and transparent government, as well as crowdsourcing initiatives and citizen engagement, can add another complicated management burden, given the heightened public accountability that this entails.

7. *The personality factor.* Local government organizations employ a wide variety of public safety, engineers, planning, natural resources, economic development, and information technology (IT) professionals. Supplementing this array of disciplines is the idea that there are three different perspectives about how to implement the GIS, which is discussed in Chapter 3: (1) the *technical guru's perspective* who believes that GIS implementation is a technical straightforward process, (2) the *management perspective* that believes that successful GIS implementation is about rational management and technical competence, and (3) the *social and cultural perspective* that empathizes the cultural and social dynamic of implementation.

8. *Funding.* Anyone who was working during the economic downturn of 2008 will understand that some of the most precious factors in deploying and maintaining an enterprise GIS are the budget and funding mechanisms. Whether it comes from enterprise funds, operational funds, or capital funds, the funding for GIS requires constant management and monitoring. Though departmental payback models that fund the GIS are often cumbersome to manage and monitor, the budgetary concerns further politicize GIS implementation and must be taken seriously. Additionally, an organization should consider grants and educate itself on the art of writing grant applications.

9. *Workflow complexity.* Standard operating procedures (SOPs) are developed to educate staff and simplify complex but repeatable procedures and workflows. SOPs are part of the GIS landscape and require a detailed understanding of business tasks.

10. *Mobility, technology, and architecture.* The deployment of a true enterprise GIS requires understanding and deploying the correct hardware, networking, software, and operating systems. Network communications, software architecture, and data security must be addressed. In the modern world, *office-to-field* and *field-to-office* capabilities are paramount to GIS success. Understanding mobile software and the connective potential it brings to GIS technology is critical to maximizing efficiency.

11. *The nonlinear history of GIS deployment in local government is a barrier to solid GIS governance.* The adoption of GIS in local government originated with a haphazard, unstructured, and decentralized approach. One could call this *organic growth.* Though organic growth can be a good thing in other contexts, I am using this term to refer to a situation where, for example, the planning department, the public works department, and the police department are all invested in GIS software that is specific to their needs without consideration for the enterprise as a whole. Therefore, the use of GIS was naturally decentralized and left each department to essentially *do its own thing.* This, for a whole variety of economic and logistical reasons, is a problem. After the GIS began to evolve with local governments, a more centralized GIS model for GIS governance was created, whereby one department of an organization is accountable for all the GIS management and oversight in the organization at large. This model, however, never really caught on as it tended to stifle creativity and the enterprising use of GIS.

12. *The need for data, open data, and open government.* Digital GIS data comes with many complexities, including the art of data conversion, the multiple ways to create accurate and reliable data, and methods for data sharing and publishing. There are many state, Federal, and local data sources that are accompanied by data standards. Open government and open data support techniques that promote freedom of information. All add more complexity to governing GIS within local government.

Now, we approach a discussion of GIS management with the phrase *hybrid governance model.* The hybrid model follows an intuitive logic that says, "Just as organizations themselves mature at different rates, their GIS governance or management models must also mature at different rates." The young GIS community started out with a decentralized approach, and then experimented with a centralized solution, before finally developing the hybrid governance model. Though the hybrid governance model does offer unparalleled flexibility, at this time, there is no *garden variety* governance model that is currently in use. Organizations must determine what works best for the unique circumstances that they face.

6.7 Three GIS Governance Models

A GIS governance model is an organizational structure and process that facilitates GIS technology growth and use. It follows that a sound governance model is essential for effective technology diffusion and ongoing management. The appropriate governance model becomes even more important when considering multidepartmental and shared GIS resources. Figure 6.4 describes the simple definition of the different governance models being used in local government organizations. In addition to the three governance models that are described in Figure 6.4, we will discuss an approach that is currently trending and includes a shared regional governance model of GIS.

The three primary governance models used to implement enterprise-wide GIS within local government organizations throughout the United States are the *centralized, decentralized,* and *hybrid models.* Even though the next few chapters may be an oversimplification of what is going on in the real world,

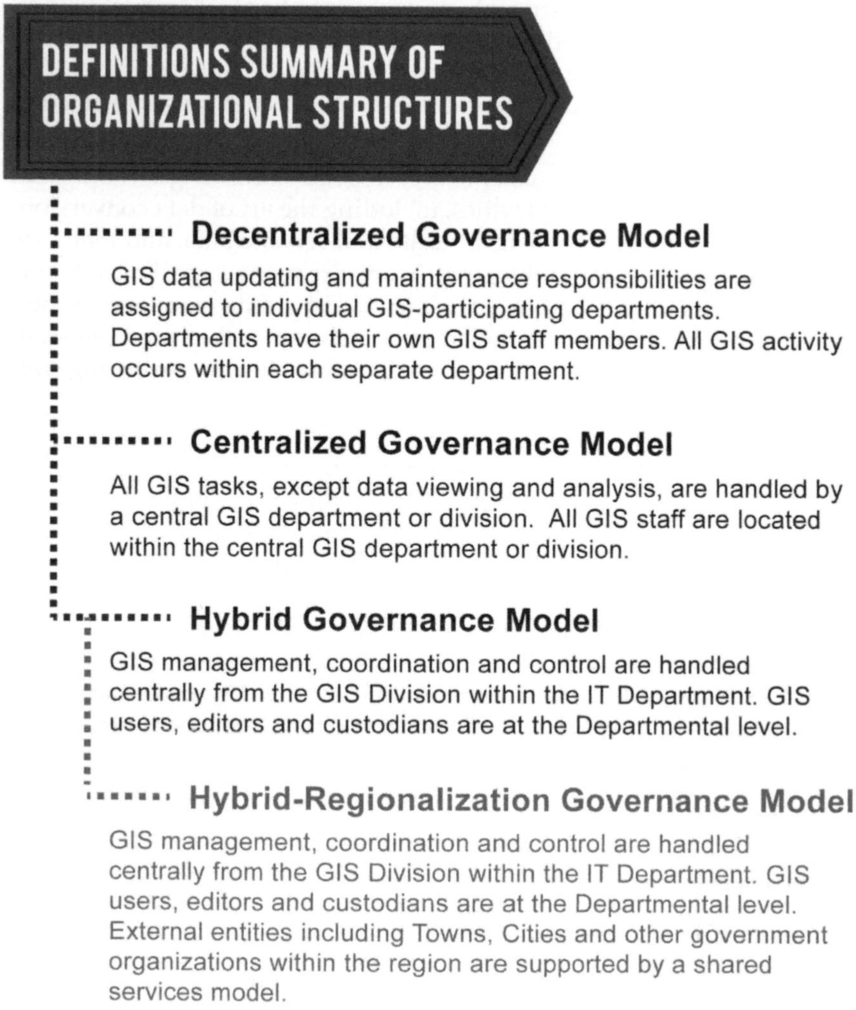

DEFINITIONS SUMMARY OF ORGANIZATIONAL STRUCTURES

Decentralized Governance Model

GIS data updating and maintenance responsibilities are assigned to individual GIS-participating departments. Departments have their own GIS staff members. All GIS activity occurs within each separate department.

Centralized Governance Model

All GIS tasks, except data viewing and analysis, are handled by a central GIS department or division. All GIS staff are located within the central GIS department or division.

Hybrid Governance Model

GIS management, coordination and control are handled centrally from the GIS Division within the IT Department. GIS users, editors and custodians are at the Departmental level.

Hybrid-Regionalization Governance Model

GIS management, coordination and control are handled centrally from the GIS Division within the IT Department. GIS users, editors and custodians are at the Departmental level. External entities including Towns, Cities and other government organizations within the region are supported by a shared services model.

FIGURE 6.4
Definitions of GIS governance models.

an examination of the world in all of its complexity would be too overwhelming for the current context. In view of this, I consider a simplistic approach to be far more manageable for our current needs.

6.7.1 Governance Model #1: Centralized Governance Structure

A centralized organizational structure maintains a central department or division that is responsible for all GIS services. In this type of structure, the GIS often has its own dedicated department or is a division of an IT or technology services department. The GIS division employs a cadre of management, analysts, technicians, and programmers who are tasked with overseeing hardware, software, application development, technical planning, and training. Data are created and maintained by this group. All other participants are characterized as end users, with only the capability to view, query, and analyze spatial data. Figure 6.5 graphically illustrates what a centralized model would look like.

Departments or business units within a centralized structure use the data for day-to-day operations and analysis. Feedback is channeled through the chain of command to the GIS information officer or coordinator with oversight coming from a steering committee and end-user groups. Bureaucracy and duplication of effort are minimized since there is a central command and control and a single budget source. Often, GIS functions are split into teams, with each team being held responsible for specific functions and the requests for services.

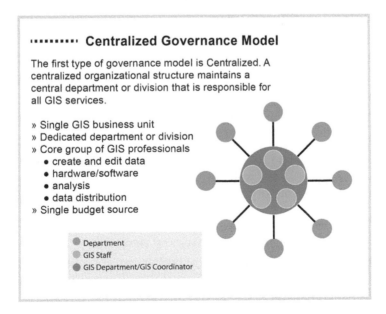

········· Centralized Governance Model

The first type of governance model is Centralized. A centralized organizational structure maintains a central department or division that is responsible for all GIS services.

» Single GIS business unit
» Dedicated department or division
» Core group of GIS professionals
  • create and edit data
  • hardware/software
  • analysis
  • data distribution
» Single budget source

  ● Department
  ● GIS Staff
  ● GIS Department/GIS Coordinator

FIGURE 6.5
Centralized GIS governance model.

My colleague and business partner, Mr. Curt Hinton, offered an excellent explanation for the centralized model:

> "This model can be compared to the military model or the waterworks model. The end user of the service relies on the central GIS business unit to provide clean GIS information. The end user just has to turn on the faucet and out flows the GIS information. He or she does not need to be aware of the effort or processes that produce the information; similarly, a person at the end of the water faucet does not have to worry about the infrastructure and management process that is required to provide clean drinking water. The centralized model is very efficient and, as such, is typically utilized by single departments, large government agencies, the military, and business corporations."

When a well-planned centralized GIS organizational structure is implemented, the government can expect the following:

- Clearly defined roles from a central chain and command.
- Standard software and maintenance procedures.
- Shared overhead costs.
- Decisive and straightforward direction.
- Solutions to operational problems that are implemented from the top down.
- Greater operational efficiency for staff throughout the organization.
- Reduction in data duplication.
- Many integration opportunities with other business systems.
- Central access point for data sharing.
- Team-based processes in which critical functions are beyond one-person deep.
- Spatial information maintenance that improves because users are well trained and devoted to specialized tasks.

In a centralized model, all GIS activity is placed in one group of experts. For this group to hold the weight of all of those GIS tasks, the governing authorities must be wary of the following issues:

- Inflexible decision making.
- Rigidified maintenance procedures and standards.
- Poorly funded implementations or budget cuts disrupting the whole system. (All the eggs are in one basket.)
- Lack of end-user input and design in the planning process.

- Smaller agencies may not know what to ask for from the central agency since they do not have any GIS experience.
- Poor centralized leadership or direction, with the top-down design, could lead to undesired results.

The major strength of the centralized model is a well-structured, well-defined, and highly efficient universal GIS system. The Achilles heel of a centralized model is its tendency to inadvertently create a GIS landscape that is too rigid or inflexible for the needs of stakeholders. It must be said that many local government organizations do not implement a centralized GIS governance model.

6.7.2 Governance Model #2: Decentralized Governance Model

The second type of GIS governance model is decentralized. As the name implies, a decentralized organizational structure divides GIS responsibilities throughout various departments. Decentralized organizational structures may still have a GIS division, operating independently or under the jurisdiction of another department, but embrace and encourage the use and maintenance of GIS throughout the organization.

Such an approach divides system and data maintenance between the GIS division and departmental end users. During their course of daily business, users update an enterprise database. All users share responsibility for maintaining the GIS, and users within each department maintain specific data according to their roles and responsibilities. This type of organizational structure enables the GIS division to focus on hardware and software maintenance, data exchange and distribution, application/data design and development, user training and support, community extension, and technology innovation, instead of devoting time to the creation and maintenance of data. Figure 6.6 graphically illustrates what a decentralized model would look like.

A decentralized GIS governance model offers the following advantages:

- Ability for departments to guide GIS activity independently from organizational initiatives.
- Bottom-up decision making.
- Line departments that are more sensitive to user needs since they are in close proximity to the developers.
- Clear lines of responsibility within the department.
- Facilitation of multitasking.
- Multiple funding sources for large projects and initiatives.
- Shared resources and costs between two departments or subdivisions.
- Willingness for staff to help each other.

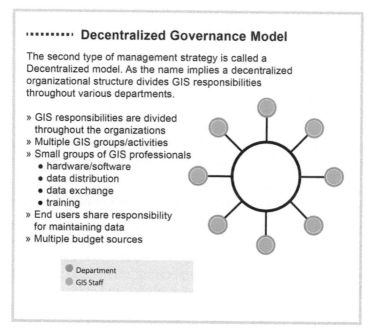

FIGURE 6.6
Decentralized GIS governance model.

A decentralized GIS governance model also comes with its own set of problems:

- Extremely strong communications, paperwork, and bureaucracy are necessary to forge agreements between multiple departments.
- Redundant roles and functions exist between departments.
- Guided by individuals rather than by teams.
- Multiple GIS and applications.
- Databases and skills that are often fragmented throughout the enterprise.
- Overhead costs that are not shared and often much higher.
  - Redundant effort in multiple departments.
  - Multiple copies of data being edited and stored in several locations.
- Difficulty in standardizing software.
- Poor data sharing and isolated databases.
- Staff wearing multiple hats and sacrificing GIS competency to day-to-day departmental operations or tasks unrelated to the GIS field.
- Staff competing with each other for funding or recognition instead of working together.

This model is often deployed by governments that are not centrally located and lack a strong technological competency. Smaller governments and those that are new to GIS technology often deploy the decentralized model.

Small jurisdictions or single departments that have a low volume of GIS work depend on this model, especially when workers have to multitask with departmental operational duties. This model also has a lower startup cost for the departments and smaller jurisdictions that make it more attractive for first-time users.

6.7.3 Governance Model #3: Hybrid Governance Model

Finally, many local governments utilize a hybrid GIS organizational model: an approach that incorporates the benefits of each of the centralized and decentralized organizational models. Think of this as a happy medium, the best of both worlds. Figure 6.7 graphically illustrates what a hybrid model would look like.

A hybrid GIS governance model uses dual accountability along functional lines. When successfully implemented, the hybrid model can benefit organizations in many ways:

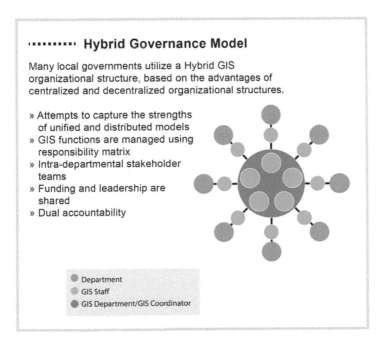

FIGURE 6.7
Hybrid GIS governance model.

- Shared costs including the following:
  - Database management and maintenance.
  - Network and server resources.
- Highly specialized GIS staff.
- Improved efficiency.
- Integrated multidepartmental solutions can be implemented.
- Central data warehouse.
- Team-based processes. (Critical functions are no longer one-person deep.)
- Improved data quality.
- The departmental ownership of relevant data sets is maintained.
- Automated validation routines.
- Real-time distribution of data.
- Improved end-user support. (Feedback from users is immediate since each team sits in close proximity to the work. They can hear and see firsthand what needs to be fixed.)

If a hybrid GIS governance model is not implemented successfully, there can be a number of problems including the following:

- Roles are not clearly defined, making expectations unclear.
- Unnecessary bureaucracy from too many standards or too many agreements and negotiations.
- *No clear direction from leadership.* Stakeholders end up setting their own priorities and looking out for their own needs.
- *Insufficient funding.* Critical functions could be cut by a single department, hurting the remainder of the enterprise.
- *Smaller departments with small staffs may be left out of the planning process and miss out on opportunities to participate.*

The major benefits from the hybrid model are derived from its flexibility. Stakeholders actively participate in the design and project-planning stages and work together while dividing and sharing the GIS functions. The GIS central body is responsible for the overall professional direction, career development, GIS system architecture, applications, license pools, and delegating project work. The intradepartmental stakeholder teams are responsible for data capture, data edits, quality control, and the cartographic output. Stakeholders pool resources and cross-train team members from different departments. Flexibility and departmental expertise are ensured, since the stakeholder teams work within the departmental structure on specific end-user functions. Redundancy is reduced, since there is a central command structure that is made up of a GIS coordinator and key GIS technical staff. If funding or leadership is lacking in a single department, then the other departments compensate. Smaller departments stay involved because they have an equal share in the decision-making process and are supported by intradepartmental teams.

The chief risk associated with the hybrid model relates to the potential disagreements among participants with regards to roles and responsibilities. An overabundance of formal agreements and meetings will confuse the decision makers and hamper productivity. An unclear overall direction can ensue. People who may not understand the system adequately will make up their own systems, just to be functional.

Strong communications, GIS knowledge, and leadership are required for the hybrid model to run smoothly. If these qualities are lacking, a hybrid model may devolve into the decentralized model, and redundant processes will emerge.

6.8 New Trending Governance Model: Regionalized Governance Model

A regionalized governance model is based on *shared services*. In this context, shared services can be defined as two or more local government authorities that plan; employ staff; undertake management, business, and/or regulatory activities; deliver and/or maintain infrastructure; or provide services to their communities in a joint manner. Such collaborative activities can be conducted in a variety of ways, ranging from simple written agreements (such as an exchange of letters) through loosely structured regional organizations of councils and other more formal entities to jointly owned companies with independent boards. A regionalized model can often be an extension to a local government centralized or decentralized governance. It essentially incorporates other municipalities into a shared services model. Figure 6.8 graphically illustrates what a regional model would look like.

6.9 Adding Functional Teams to the Mix of Governance Models

A steering committee is established to oversee all GIS activities within the organization. For that reason, one would presume that the centralized and hybrid model would always use a steering committee, whereas the decentralized model would not require such enterprise structure. Essentially, the GIS coordinator reports to the GIS steering committee. However, if the

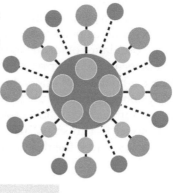

·········· Hybrid & Regionalization Governance Model

Many local governments utilize a Hybrid GIS
organizational structure that supports a regionalization of GIS.
It has the advantages of a centralized and decentralized model.

» Attempts to capture the strengths
of unified and distributed models
» GIS functions are managed using
responsibility matrix
» Intra-departmental stakeholder
teams
» Funding and leadership are
shared
» Dual accountability

● Department
● GIS Staff
● GIS Department/GISCoordinator/External Organization

FIGURE 6.8
Regionalization of GIS governance.

organization is large, it may require functional teams to be established to oversee key issues that are related to that specific topic or functional group. Different types of functional GIS groups have been established in many organizations including, but not limited to, the following:

- *Utility functional GIS team*—This team will focus on needs that pertain to departments whose primary focus is the infrastructure, with an emphasis on the services that are typically delivered through traditional utility and engineering functions.

- *Public safety functional GIS team*—This team will focus on needs that pertain to departments whose primary focus is the safety of the community.

- *Human services functional GIS team*—This team will focus on needs that pertain to departments whose primary focus is the delivery of interpersonal services.

- *Land management functional GIS team*—This team focuses on land-based GIS, including the parcel fabric zoning, land use, and political districts.

The lead GIS staff person assigned to a functional team should become an expert on the team's functions. For example, a GIS staff person assigned to the public safety and welfare team would make himself or herself intimately

FIGURE 6.9
Functional teams.

familiar with the details on how each public safety and welfare department operates from the ground level up and how GIS can benefit from that department. Team members are required to contribute to the overall goals of the team and satisfy the needs of their department. Each functional team should meet at set intervals to discuss how best to carry out the overall GIS needs of their team. Functional team members should work toward common goals such as developing and maintaining the needed data layers, special projects, and other needs. The necessary teamwork mentality will only function if supported by management. Figure 6.9 illustrates the possible functional teams within a local government organization.

6.10 Departmental Accountability, Best Business Practices, Executive Champion, GIS Technical Committee, GIS Steering Committee, Subject Matter Experts, and GIS User Groups

The term dual accountability describes the dual roles of some individuals within the organization. More specifically, dual accountability implies that a department should satisfy the goals of their department while simultaneously meeting the goals of an organization at large. A GIS analyst in the Department of Public Works has two bosses: (1) the public works director and (2) the GIS coordinator. The GIS analyst has dual accountability as he or she reports directly to the department director as well as the GIS coordinator. It is important to have a clear line of responsibility and an understanding of dual accountability. Some other terms that are important in local government GIS initiatives are as follows:

- Best GIS business practices

  Best business practices (BBPs) are a method, practice, or technique that consistently shows that it is the most effective way of performing that specific task. It is the standard practice for performing a task. It could be the benchmark for quality practices. Often, BBPs are accredited standards like Esri's Local Government Information Model.

- Executive GIS champion

  Communities with strong GIS implementation records generally have an individual at the highest levels of the organization, who is able to bring parties together in order to develop a shared vision of the benefits that a GIS can bring to the community. An executive champion is typically a mayor, a city manager, a chief information officer, or a council person.

- GIS technical committee

  The technical committee typically shapes the technical direction and establishes the technical policy of the GIS initiative. Most local government organizations fold this duty into the GIS steering committee's duties. After all, you can committee things to death.

- GIS steering committee

  The steering committee typically shapes the funding for and the direction and policy of the GIS initiative. Nearly every local government organization utilizes a GIS steering committee. Most steering committees are made up of department directors. Look for the old adage "A camel is a horse designed by a committee."

- Subject matter experts

  Subject matter experts (SMEs), as the name implies, are the individuals who are recognized as an expert in a specific field. Often, GIS governance models discuss the GIS SME existing within a specific department. For example, the police department would have an SME in a crime analyst.

- GIS user group

  A GIS user group is a cohort of stakeholders, generally grouped by geographic location, who share information and compare experiences with GIS technology for the benefit of all members.

6.11 An Evaluation of the Different Governance Models

Figure 6.10 ranks the potential benefits for each of the governance models that are discussed in this chapter. Additionally, it ranks the typical challenges

GOVERNANCE MODEL COMPARISON			
Potential Benefits to the Organization	**Centralized Model**	**Decentralized Model**	**Hybrid Model**
• Clearly defined roles reducing conflicts of confusion about service • Enterprise level direction and goals • Central chain of command (top-down solutions) • Clear and straightforward (I need a map) • Quick and fully informed decision making • Predictable format	👍👍	👍	👍👍👍
• Shared costs reduced • Database management and maintenance • Network and server resources • Highly specialized GIS staff	👍👍👍	👍	👍👍👍
• Achieving stakeholder needs • Departments contribute GIS input and resources • Sensitive to department and user needs	👍	👍👍	👍👍
• Reduction duplication • Data (multiple copies of data) • Effort (data creation and maintenance) • Project initiatives and expenses	👍👍👍	👍	👍👍👍
• Improved data sharing/integration with other business systems • Enterprise systems • Multi-departmental solutions • Central access point	👍👍	👍	👍👍👍
• Institutional Legacy • Team-based processes • Cross-training of employees • Fail-safe critical GIS functions and tasks (beyond one person deep)	👍👍👍	👍👍	👍👍👍
• Clear departmental expectations • Responsibilities • Participation • End-user knowledge	👍	👍👍👍	👍👍
Expected Challenges to the Organization			
• Potential for too many standards (formal agreements proliferate) • Too many meetings and committees • May require extensive negotiations • Difficult to understand	👎	👎👎👎	👎👎
• Potential for too rigid standards (more time is devoted to following standards and the letter of the law and less to the original purpose of the program)	👎👎👎	👎	👎
• Funding risks (if funding is suddenly lost) • All the eggs in one basket	👎👎👎	👎👎	👎
• Exclusion of Smaller Departments (if everyone is not equal • Funding • Service • Technology	👎	👎👎👎	👎
• Risk for departmental system isolation (everyone does their own thing) • Solo initiatives • Lack enterprise cooperation • Risk of pull outs or refusals to participate	👎	👎👎👎	👎

FIGURE 6.10
Evaluation of governance models.

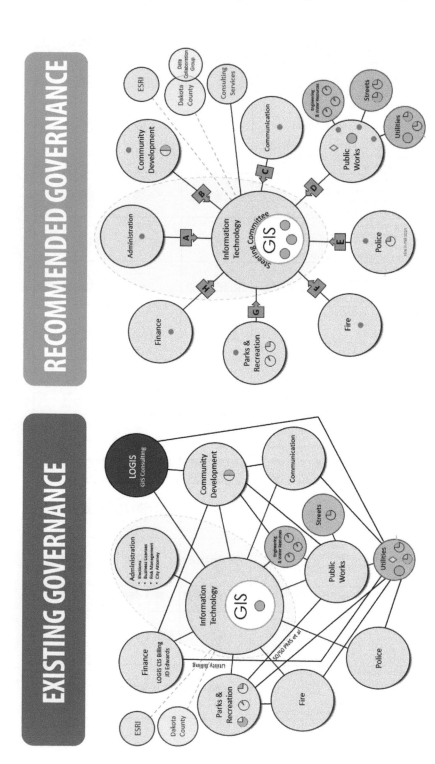

FIGURE 6.11

City A—existing and future GIS governance model.

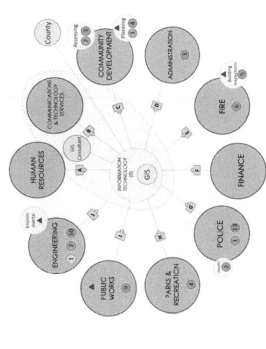

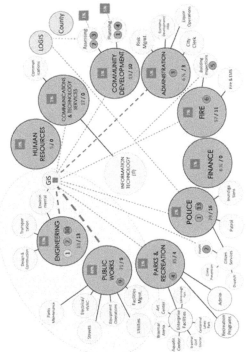

FIGURE 6.12

City B—existing and future GIS governance model.

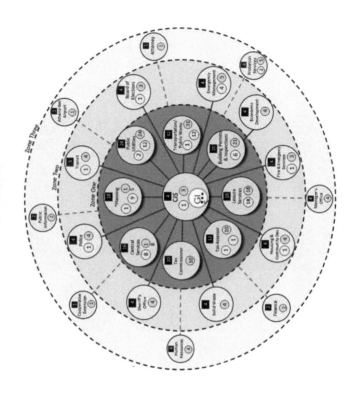

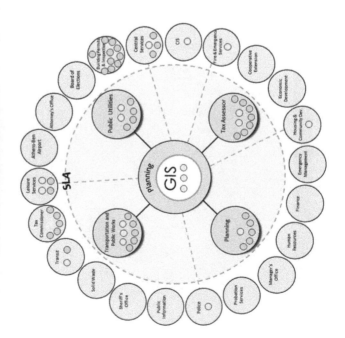

FIGURE 6.13
County C—existing and future GIS governance model.

that are faced when implementing a governance structure. Thumbs-up and thumbs-down icons are used to represent how each model performs for each element.

6.12 Case Study Discussion

Since 2010, there has been a significant upturn in the demand for evaluating existing GIS governance within local government organizations. This has led to a renewed effort in evaluating, planning, designing, and explaining governance options to towns, cities, and counties. Each organization is different, and each organization is at a different phase of GIS maturity. Figures 6.11 through 6.13 illustrate three local government organizations that all had existing centralized GIS governance characteristics. Each organization is moving toward the more popular hybrid model that offers sustainability with a corresponding increase in the number of GIS users.

7

GIS Training, Education, and Knowledge Transfer

Educating the mind without educating the heart is no education at all.

Aristotle

7.1 Introduction

We need to start looking at GIS education, training, and knowledge transfer from a completely different perspective. We should consider what the modern-day GIS coordinator needs. Do not let the flashiness of software solutions lead our line of thinking. We often focus in on the training and education surrounding the software itself and forget all about the *other stuff*. The other stuff includes people management, governance, creative thinking, communication, training and educating the organization, presenting, selling, aligning, developing performance measures, writing grant applications, and developing service level agreements (SLAs), just to name a few.

7.2 GIS Timeline—GIS Management and Training and Education

Before I discuss the future of GIS training and education, let me explain what has happened in the local government GIS industry. The timeline in Figure 7.1 graphically and simplistically depicts the evolution of GIS with special emphasis on training, education, and knowledge transfer practices.

7.2.1 Period 1: Big Bang to 1970s—The Geographer and Cartographer

My description of this period up to 1970 is more tongue in cheek than anything else and is intended to summarize the history of GIS from time *immemorial*, that

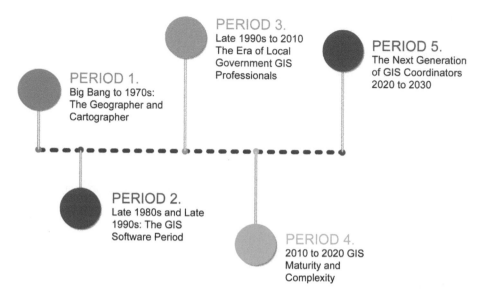

FIGURE 7.1
GIS timeline.

is to say, a period beyond the reach of memory and record. Well, at least it is, as it relates to this book. We should say that this period started with cave paintings, stone tablet carvings, and ancient brush parchment maps. In our more recent history, advances in electronics, cartography, automated drafting, and map interpretation, as well as geographic theory, initiated what we could call *old-school* training and education. Arguably, we could say that, before the 1970s, GIS training could have been measured by cartographic principles, map projections, urban and rural geography, physical geography, and human settlement. My point here is that the GIS really began in the 1970s. Before then, our local governments had little idea as to the use and benefit of the tools that are afforded by the GIS.

7.2.2 Period 2: Late 1980s to Late 1990s—The GIS Software Period

Yes, I know that the origins of GIS can be traced back to the early 1960s. However, I only entered the scene in late 1987, so I can only tell you what I have observed. Remember that 99% of what you know is secondhand information, but at the moment, I am using that rare 1%. The attitude that defined this period could be expressed as, "Let's learn this new GIS software as fast as we can, and it should translate into a career with a potential for rapid promotion."

Initially, GIS software was in the driver's seat. Twenty years ago, many university graduates entered the marketplace and became GIS specialists, GIS analysts, GIS managers, or GIS coordinators for our towns, cities, and counties, all because they learned this new GIS software. Ask any GIS coordinator over the age of 40 what his or her academic background is, and I suspect that most will not have a formal GIS background or even an education in geography. They came from the school of engineering, planning, natural resources, forestry, parks and recreation, humanities, architecture, and landscape architecture among others.

The majority of GIS professionals in local government learned GIS by becoming proficient with the software through experiential learning rather than a formal GIS-directed education at their university. By the turn of the millennium, we had thousands of local government GIS professionals with expertise in GIS software but with gaps in other areas. This period was marked by GIS software training that focused on Unix Server GIS–spatial analysis–raster and vector analysis, spatial modeling, overlaying, and statistics.

7.2.3 Period 3: Late 1990s to 2010—The Era of Local Government GIS Professionals

At the same time that GIS-specific employment gathered steam in local government organizations, GIS became far more information technology (IT)–centric and more complex to manage and maintain. At this point, GIS demanded more than just software expertise. GIS governance models and management styles reared their ugly heads. During this period, the demand for GIS strategic implementation planning quadrupled. Essentially, we spent this period finding out the extent of the reach of GIS solutions. Local government demanded answers to the newly identified tactical, technical, strategic, logistical, and political problems of GIS. Ultimately, the period between 2000 and 2010 was marked by a widespread acceptance of GIS as an enterprise business solution for local government.

It is worth noting that the changes in Esri training corresponded to both industry and larger societal trends. If you performed a historical review of Esri training and education offerings, you will see how it has evolved into something far more comprehensive than mere software training. It includes education workshops on nonsoftware-related issues, such as the following:

- Workflows and analysis
- Geospatial intelligence
- Map projections and coordinate systems
- Metadata
- Cartographic design
- Turning data into information
- Good and best business practices
- Standard operating procedures (SOPs)

Esri recognized the demand within the local government market for training and education about the critical components of GIS. Not just the actual software itself.

7.2.4 Period 4: 2010 to 2020—GIS Maturity and Complexity

Local government may have started to hit the proverbial wall when it comes to GIS. We now recognize that implementing a truly enterprise and sustainable GIS is more difficult than first thought. It would seem that GIS has more moving parts than an advanced LEGO set. GIS is not easy because it has so many components. My formula for success in Chapter 2 demonstrates this very fact.

I believe that the next five years will highlight the notion that GIS has matured into a highly complex and wide-ranging system. Local government organizations now have to address issues and challenges that were never before considered, namely, the governance and management of GIS along with the traits that make up an ideal GIS coordinator. If we look closely at what is required of a GIS coordinator, we can shape the design and future of our digital classroom.

GIS moved from the desktop to the cloud overnight. Big data analytics and dashboarding of information came into vogue. Demand rose for simple tools to tell geographic stories and easy field data collection and maintenance. Esri refers to this as the *Maps and Apps* phase.

7.2.5 Period 5: 2020 to 2030—The Next Generation of GIS Coordinators

We are fast approaching the era of the geographic information officer (GIO). Elected officials, city managers, and senior staff are already recognizing the potential place for a GIO within the organization. Obviously, this depends on the size and complexity of an organization, but GIOs will play significant and positive roles in a new wave of enterprise GIS and the education and knowledge transfer that are required to sustain this type of deployment. Figure 7.1 simply illustrates my ideas about what has happened to GIS management, training, education, and knowledge transfer since the beginning and into the future.

7.3 Type of GIS Users in Local Government

At this point, I want to explain that when we develop GIS strategic implementation plans for a local government organization, we always create a comprehensive and department-specific GIS software training program. We offer training and education based on GIS software demand. We need to do this because we have *four tiers or types of GIS users* within local government that use a variety of GIS applications. Figure 7.2 illustrates the tiers and/or types of GIS users within local government.

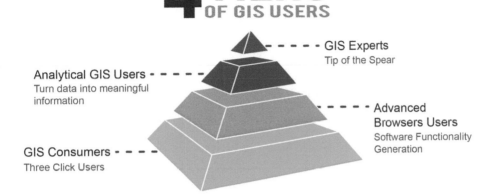

FIGURE 7.2
Four tiers of GIS users.

1. *GIS experts*—Tip of the spear
2. *Analytical GIS users*—Turn data into meaningful information
3. *Advanced browsers users*—Software functionality generation
4. *GIS consumers*—Three-click users

The type and usage of GIS software in local government is diverse and extensive. Looking at the complexity of geospatial needs on a department-by-department basis illustrates the magnitude of this task. The following list is an example of typical local government departments:

- Airport
- Attorney
- Supervisor of elections and regulations
- Building permits and inspections
- Central services
- IT
- Cooperative extension
- Economic development
- Emergency management
- Finance
- Fire and emergency services
- Housing and community development
- Human resources

- Parks and recreation
- Manager's office
- Organization development
- Planning
- Police
- Probation services
- Public information office
- Public utilities
- Sheriff
- Solid waste
- Tax assessor's office
- Tax commissioner
- Transit
- Transportation
- Public works

7.4 Esri-Based GIS Software Solutions

The following are brief descriptions of Esri software solutions that are commonly found in use in local government.

ArcGIS for Desktop

- *Basic*

  The Basic license level for ArcGIS for Desktop provides tools to interact with digital data visually and create maps. Further functionality includes the ability to view Computer-aided design (CAD) data or satellite images and the means to generate reports and charts from spatial data.

- *Standard*

  The Standard licensing option for ArcGIS for Desktop includes all Basic functionality, as well as tools for advanced data management. This level gives users complete GIS data-editing capabilities, the option for disconnected editing, and automated quality control tools. In ArcGIS for Desktop, standard, raster-to-vector conversions can be processed, and spatial data can be created from scanned paper maps.

- *Advanced*

  Advanced ArcGIS for Desktop, introduces users to advanced analysis, high-end cartography, and extensive database management while maintaining all the functionality from Basic and Standard license levels. This licensing option further promotes GIS data modeling, advanced data translation and creation, and advanced feature manipulation and processing.

Extensions

- *ArcGIS Spatial Analyst*—Derives answers from your data using advanced spatial analysis.
- *ArcGIS Geostatistical Analyst*—Uses advanced statistical tools to investigate your data.
- *ArcGIS Network Analyst*—Performs sophisticated routing, closest facility, and service area analysis.
- *ArcGIS Infographics Add-In*—Gets facts and demographics about areas with clean, clear infographics.
- *ArcGIS Data Reviewer*—Automates, simplifies, and improves data quality control management.
- *ArcGIS Workflow Manager*—Better manages GIS workflows and tasks.
- *ArcGIS Pro*

ArcGIS Pro reinvents desktop GIS. It is a central part of the ArcGIS platform. This brand-new 64-bit desktop application lets users render and process data faster than ever. Designing and editing in 2D and 3D, working with multiple displays and layouts, and publishing finished Web maps directly to ArcGIS Online or Portal for ArcGIS, connects users throughout the organization or the world.

- *ArcGIS Online*

  ArcGIS Online allows organizations to use, create, and share maps, scenes, apps, layers, analytics, and data. It gives access to ready-to-use apps, and Esri's secure cloud, where the organization can add items and publish Web layers. Because ArcGIS Online is an integral part of the ArcGIS system, it can be used to extend the capabilities of ArcGIS for Desktop. ArcGIS Online includes a wide range of apps that will interact with the maps and data within the organization, including apps for the field, office, and community and app builders for the Intranet and Internet. The following apps are part of Esri's ArcGIS Online suite:

- *Intranet Map Viewer*

  The Intranet Map Viewer is an intuitive solution for viewing, querying, and analyzing data. It is a tool that offers new and innovative decision support and a solution that will improve departmental operations.

- *Internet Map Viewer*

  The Internet Map Viewer is an intuitive solution for the public to view, query, and analyze selected digital data.

- *Mobile Field Data Collector*

  The Mobile Field Data Collector is an easy-to-use map-centric solution to inventory, update, and maintain digital data. It takes advantage of smartphones, iPads, and tablets. The task of collecting data in the field and turning it into meaningful information in the office has never been easier.

- *Operations Dashboard*

  An Operations Dashboard is a unique and interactive command center that dynamically monitors information in a dashboard environment. This is a new solution to monitor the status and performance of specific information that is used by departments. It easily tracks field workers and keeps an up-to-date record of existing information about departmental assets.

- *Story maps*

  A solution that offers a department improved community awareness and engagement through geospatial story maps. This is a solution that offers an exciting and visually pleasing way of engaging citizens.

- *Dynamic Database Integration*

  The Dynamic Database Integration is a solution that offers a real and present opportunity to integrate a department's existing systems data into the Intranet Map Viewer.

Figure 7.3 lays out a graphic depiction of the departmental use of the Esri suite of software. This matrix defines the Esri products on the y-axis, and county government departments on the x-axis. This is a simple way to understand existing and future GIS usage and need. I deploy the graphic in Figure 7.3 to explain the complexity of each department's needs. Like all good starting points, it is fairly simplistic.

7.5 Defining GIS Training, Education, and Knowledge Transfer

GIS education, training, and knowledge transfer describes the process of gaining knowledge and developing skills. It allows staff to efficiently

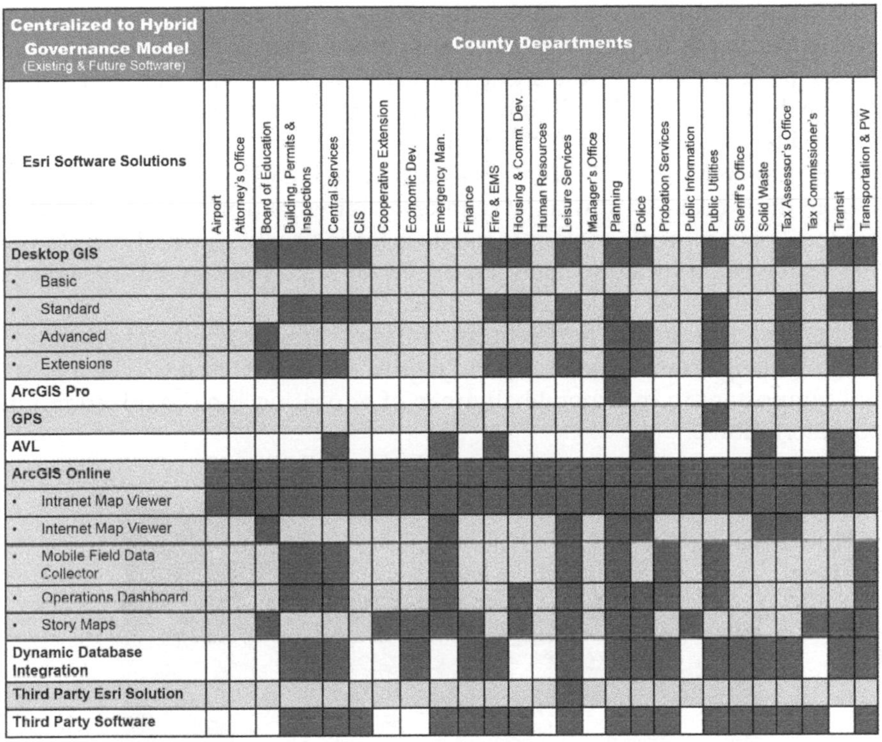

Centralized to Hybrid Governance Model (Existing & Future Software) — **County Departments**

Esri Software Solutions across County Departments: Airport, Attorney's Office, Board of Education, Building, Permits & Inspections, Central Services, GIS, Cooperative Extension, Economic Dev., Emergency Man., Finance, Fire & EMS, Housing & Comm. Dev., Human Resources, Leisure Services, Manager's Office, Planning, Police, Probation Services, Public Information, Public Utilities, Sheriff's Office, Solid Waste, Tax Assessor's Office, Tax Commissioner's, Transit, Transportation & PW

Esri Software Solutions rows:

Desktop GIS
- Basic
- Standard
- Advanced
- Extensions

ArcGIS Pro

GPS

AVL

ArcGIS Online
- Intranet Map Viewer
- Internet Map Viewer
- Mobile Field Data Collector
- Operations Dashboard
- Story Maps

Dynamic Database Integration

Third Party Esri Solution

Third Party Software

FIGURE 7.3
Departmental matrix of potential Esri software use.

reason, deploy good judgment, and supplement their personal chests of knowledge. Education, training, and knowledge transfer directly influences the success and utilization of an enterprise GIS. It allows a local government to maintain its competitiveness. Education is a world-changing force, and we should be critical of the ways that it manifests at the local government level. The value of GIS education will ultimately depend on the type of GIS education tools that are deployed. It has numerous benefits that include the following:

- Changing the culture of your organization
- Reducing ignorance
- Promoting awareness
- Promoting creative and innovative thinking
- Improving performance of GIS activities
- Improving self-esteem and confidence

Training, education, and knowledge transfer will continue to change the local government technology landscape. In the immortal words of Nelson Mandela, "Education is the most powerful weapon that you can use to change the world." Sir Ken Robinson also states that "Creativity is as important as literacy and we should treat it with the same status." I believe that we need to embrace GIS education as a way to creatively and thoughtfully change the world. An aim so large takes original and innovative thinking, and local governments serve as incubators for innovative ideas. Let us have a sneak preview of things to come. Figure 7.4 shows a colleague using a drone to capture a video of inaccessible areas in Alaska.

This new technology will require software training, ethical education, and knowledge transfer solutions. As Figure 7.5 illustrates, it will be increasingly important to incorporate all areas of educating the organization into job descriptions.

FIGURE 7.4
Drone technology.

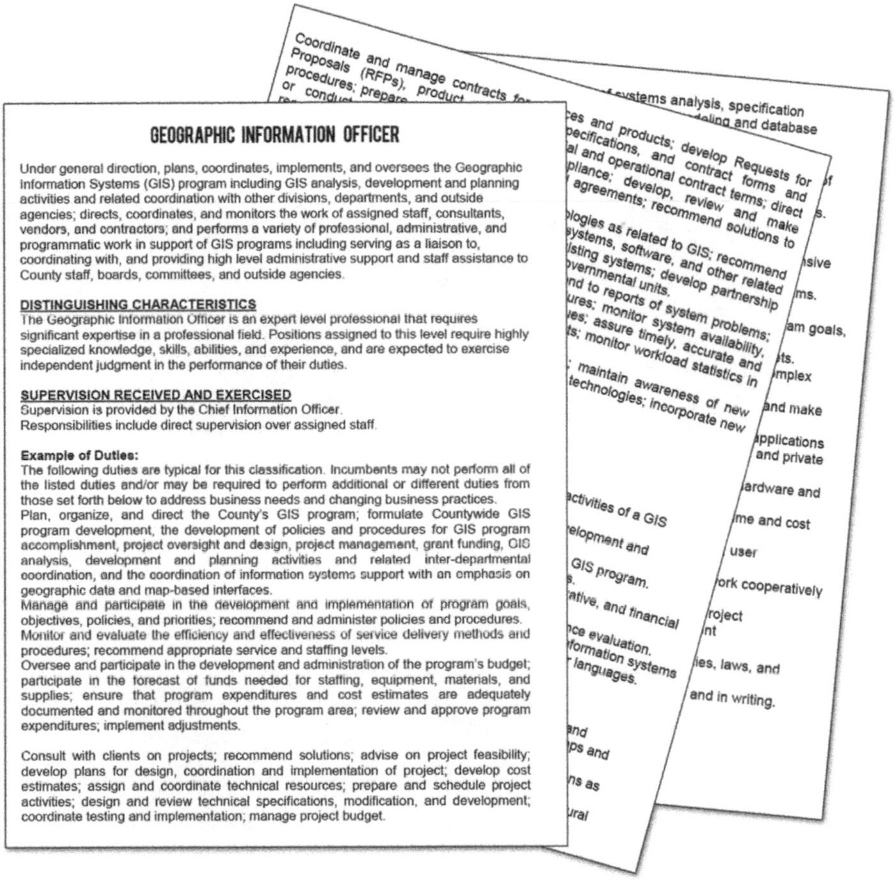

GEOGRAPHIC INFORMATION OFFICER

Under general direction, plans, coordinates, implements, and oversees the Geographic Information Systems (GIS) program including GIS analysis, development and planning activities and related coordination with other divisions, departments, and outside agencies; directs, coordinates, and monitors the work of assigned staff, consultants, vendors, and contractors; and performs a variety of professional, administrative, and programmatic work in support of GIS programs including serving as a liaison to, coordinating with, and providing high level administrative support and staff assistance to County staff, boards, committees, and outside agencies.

DISTINGUISHING CHARACTERISTICS
The Geographic Information Officer is an expert level professional that requires significant expertise in a professional field. Positions assigned to this level require highly specialized knowledge, skills, abilities, and experience, and are expected to exercise independent judgment in the performance of their duties.

SUPERVISION RECEIVED AND EXERCISED
Supervision is provided by the Chief Information Officer.
Responsibilities include direct supervision over assigned staff.

Example of Duties:
The following duties are typical for this classification. Incumbents may not perform all of the listed duties and/or may be required to perform additional or different duties from those set forth below to address business needs and changing business practices.
Plan, organize, and direct the County's GIS program; formulate Countywide GIS program development, the development of policies and procedures for GIS program accomplishment, project oversight and design, project management, grant funding, GIS analysis, development and planning activities and related inter-departmental coordination, and the coordination of information systems support with an emphasis on geographic data and map-based interfaces.
Manage and participate in the development and implementation of program goals, objectives, policies, and priorities; recommend and administer policies and procedures.
Monitor and evaluate the efficiency and effectiveness of service delivery methods and procedures; recommend appropriate service and staffing levels.
Oversee and participate in the development and administration of the program's budget; participate in the forecast of funds needed for staffing, equipment, materials, and supplies; ensure that program expenditures and cost estimates are adequately documented and monitored throughout the program area; review and approve program expenditures; implement adjustments.

Consult with clients on projects; recommend solutions; advise on project feasibility; develop plans for design, coordination and implementation of project; develop cost estimates; assign and coordinate technical resources; prepare and schedule project activities; design and review technical specifications, modification, and development; coordinate testing and implementation; manage project budget.

FIGURE 7.5
GIO job description.

Before we discuss this subject in detail, let me define what we mean by GIS training, education, and knowledge transfer.

- *GIS training*: GIS training is the action of *teaching a particular skill* or a new type of behavior. Training tends to be more formal and often includes computer technology. Examples include GIS software training, Global Positioning System (GPS) field data collection, and database design. Figure 7.6 illustrates GIS training in local government.
- *GIS education*: GIS education is the *enlightened experience* that follows systematic instruction and usually occurs in an academic setting. Education is less formal than GIS training and does not include anything but the student's presence. Examples include GIS education workshops. Figure 7.7 illustrates education in local government.

FIGURE 7.6
GIS training.

FIGURE 7.7
GIS education.

- *GIS knowledge transfer*: GIS knowledge transfer is the art of *transferring knowledge* from one part of the organization to another. This is usually accomplished in a very relaxed atmosphere. Examples include Blue Sky sessions; strengths, weaknesses, opportunities, and threats analysis; and GIS brown bag meetings, GIS steering committee meetings, GIS user group presentations, and GIS Day. Knowledge transfer seeks to organize, capture, and distribute knowledge in order to make it available to others within the organization. Figure 7.8 illustrates knowledge transfer in local government.

FIGURE 7.8
GIS knowledge transfer.

For the duration of my career, local government professionals have rated training, education, and knowledge transfer consistently as a missing component of their GIS initiative. The irony is that in my experience, training, education, and the subtle art of transferring knowledge are areas of GIS implementation that are poorly planned, unstructured, and ill-attended. Local governments badly want sustainable solutions and fail to implement them for this reason.

As GIS moves into the next generation, it is important that we develop an integrated approach to training, education, and knowledge transfer. Figure 7.9 illustrates how we should combine each initiative and literally *brand it* as an integral part of any enterprise, sustainable, and enduring GIS.

Figure 7.10 is an illustration of how local government organizations relaunch their GIS and embrace all of the educational components. The training, education, and knowledge transfer workshops in Figure 7.10 should be conducted every month for a year.

FIGURE 7.9
Integrated training, education, and knowledge transfer.

Learn How to Utilize the Full Extent of GIS

Introduction Series

Introduction to Geographic Information Systems (GIS)
What is GIS?

GIS Strategic Planning and Management
Planning, designing, and managing an enterprise and sustainable GIS

Management Series

GIS: A Profitable Initiative
Creating a business case for GIS

The Architecture of GIS
Critical GIS Components

The GIS Obstacle Course
Overcoming GIS Challenges, Barriers, and Pitfalls

Organization's Enterprise GIS: State of the Union
Organization's GIS Implementation Plan and Status Update

The Future of GIS
The Future of GIS Technology

Software Series

Exploring the World of Esri Software
Demonstrating the Esri Software Suite

Local Government Business Applications
How to effectively use GIS in Local Government operations

The Mobility of GIS: How mobile is GIS?
Using GIS in the field to collect, maintain and update your databases

GIS and Public Safety Operations
Utilizing GIS for Public Safety and Emergency Operations

GIS and the Natural Environment
Using GIS to manage parks, land, and natural resources

FIGURE 7.10
Local government education and knowledge transfer seminars.

7.6 Characteristics of a Perfect GIS Coordinator

This leaves us with the need to develop training, education, and knowledge transfer around the needs of the future GIS coordinator and, more specifically, the needs of a future workforce. Figure 7.11 offers a blueprint for things to come from the technical and managerial side. As the chief executive officer of a small, innovative company, I find it frustrating that the one thing missing in the labor pool of GIS staff is the ability to think creatively. Let us talk about what these creative characteristics are and how we can translate them into opportunities for training, education, and knowledge transfer opportunities. We have three ideas to talk about:

1. How to become an effective GIS leader
2. The modern-day skills of a GIS coordinator
3. Creative leadership

Two books by Geoff Colvin influenced my thinking: (1) *Talent Is Overrated: What Really Separates World Class Performers from Everyone Else* (2008) and (2) his follow-up book titled *Humans Are Underrated: What High Achievers Know that Brilliant Machines Never Will* (2015). I supplemented Colvin's findings with the ideas and principles that are promoted by the Center for Creative Leadership. Both books and the ideas promoted by the Center for Creative Leadership (CCL) not only support my thinking when it comes to GIS training, education, and knowledge transfer in local government, but they also add some additional facts that will help us plan for the future.

7.6.1 Idea Number One: How to Become an Effective GIS Leader

According to Colvin, "It isn't specific, innate talents or plain old hard work that will make you an incredible performer" in the world. It is something that we can all do. Colvin *nails down* what does not drive great performance. It is not experience. It is not inborn abilities. It is not intelligence and memory.

Colvin posits that *incredible performers* undertake *deliberate practices*, characterized by an activity that is specifically designed to improve performance. It does necessitate continuous feedback and is highly demanding mentally. If we take luck, time, and chance out of the equation, we are left with some key principles of deliberate practice.

The Principles of Deliberate Practice

- Designed practice.
- Qualified high repetition.
- Continuous feedback from mentor, coach, or teacher.

Characteristics of an Enterprise GIS Coordinator or Geographic Information Officer (GIO)

Great Communicator and Presenter • Educator • Problem Solving and Analytical Skills • Integrity • Competence • Ability to Delegate • Empathy • Understand the Art of Cooperation • Enthusiasm and Passion • Inspire's a Shared Vision, Goals, and Objectives • Team-Building Skills

GIS Governance and Management Skills	GIS Digital Data and Databases Expertise	Understand and Documents Procedures, Workflow, and Integration	Expert Level with Understanding the Applications of GIS Software	Expert with GIS Training, Education, and Knowledge Transfer	Understands IT Infrastructure and Architecture
STRATEGIC PLANNING Understand GIS strategic planning process and the benefits of a detailed plan. **DEVELOP VISION, GOALS, AND OBJECTIVES:** Tools including Blue Sky Session, on-line questionnaires, and departmental interviews.	**PERFORM A DIGITAL DATA ASSESSMENT** Perform a digital data assessment and review. The GIS Coordinator should be able to assess the quality, content, and completeness of every digital data layer.	**OVERSEE ENTERPRISE GIS INTEGRATION** Understands enterprise integration. The GIS Coordinator should understand the business and ROI opportunities for integrating GIS with existing legacy solutions.	**MANAGE GIS SOFTWARE LICENSES** The GIS Coordinator is ultimately responsible for managing GIS licenses and any enterprise license agreements (ELA).	**DEVELOP A FORMAL GIS TRAINING PLAN** The GIS Coordinator is ultimately responsible for creating a formal ongoing GIS training plan.	**CONDUCT STRATEGIC IT PLANNING FOR AN ENTERPRISE GIS** The GIS Coordinator is ultimately responsible for understanding the strategic technology plan.
UNDERSTAND GOVERNANCE MODELS Enforce and manage a formalized governance model.	**CREATE A COMPREHENSIVE MASTER DATA LIST WITH CUSTODIANSHIP** Develop and maintain a master data list. The GIS Coordinator should maintain an accurate and reliable Master Data List and share this with all departments.	**IDENTIFY GAPS AND OPPORTUNITIES** Identify opportunities and gaps. All gaps and opportunities for GIS integration should be identified and presented by the GIS Coordinator.	**MANAGE COMMERCIAL OFF THE SHELF (COTS) VERSES CUSTOM CODE** The GIS Coordinator is ultimately responsible for enforcing policies about the level of GIS costs verses custom code.	**CONDUCT MULTI-TIERED GIS TRAINING AND EDUCATION** The GIS Coordinator is ultimately responsible for understanding multi-tiered GIS software training.	**CREATE AN ARCHITECTURAL DESIGN FOR AN ENTERPRISE GIS** The GIS Coordinator is ultimately responsible for developing a high-level GIS architectural design.
DETAIL GIS JOB CLASSIFICATIONS AND DUTIES Understand job duties and job classifications.	**DEVELOPING METADATA STANDARDS AND ENFORCE THE CREATION OF METADATA** Keep Metadata. The creation of Metadata for every digital data layer should be encouraged and enforced by the GIS Coordinator.	**DOCUMENT AND DETAIL CREATIVE AND INNOVATIVE DEPARTMENTAL GIS USE** Create departmental solutions for access to critical data layers. The GIS Coordinator should have the skillset to plan, design, and implement departmental GIS solutions.	**ENABLE ENTERPRISE GIS ACCESSIBILITY** The GIS Coordinator is ultimately responsible for continuing to improve access to software.	**CONDUCT MOBILE GIS TRAINING** The GIS Coordinator is ultimately responsible for administering mobile software training.	**MONITOR INFRASTRUCTURE REQUIREMENTS OF AN ENTERPRISE GIS** The GIS Coordinator is ultimately responsible for documenting and detailing IT infrastructure to meet operational needs.
ENTERPRISE PROJECT MANAGEMENT Enterprise GIS project management. **COORDINATE THE ENTERPRISE** Coordinate all GIS activities, projects and protocols - Coordinate all departmental activities.	**OVERSEE DATA LAYER MAINTENANCE** The GIS Coordinator should oversee keeping critical data layers up to date. Every digital data layer should have a departmental custodian that maintains the accuracy, content, and completeness of that digital data layer.	**DEVELOP STANDARD OPERATING PROCEDURES** Develop GIS Standard Operating Procedures (SOP). The GIS Coordinator is ultimately responsible for developing enterprise and departmental Standard Operating Procedures.		**CONDUCT DEPARTMENTAL GIS TRAINING AND EDUCATION** The GIS Coordinator is ultimately responsible for administering departmental specific education.	**UNDERSTAND IT REPLACEMENT PLANNING** The GIS Coordinator is ultimately responsible for understanding the IT replacement plan.
EFFECTIVE COMMUNICATION ACROSS ALL DEPARTMENTS AND THE ENTERPRISE Communicate effectively to the GIS steering committee. Regularly and effectively present the state of the GIS using graphic rich media.		**PLAN, DESIGN, AND DEPLOY INTRANET GIS SOLUTIONS** The GIS Coordinator is ultimately responsible for planning, designing, and implementing an Intranet solution.			

(Continued)

FIGURE 7.11

Characteristics of a perfect GIS coordinator.

(Continued)

Characteristics of an Enterprise GIS Coordinator or Geographic Information Officer (GIO)

Great Communicator and Presenter • Educator • Problem Solving and Analytical Skills • Integrity • Competence • Ability to Delegate • Empathy • Understand the Art of Cooperation • Enthusiasm and Passion • Inspires a Shared Vision, Goals, and Objectives • Team-Building Skills

GIS Governance and Management Skills	GIS Digital Data and Databases Expertise	Understand and Documents Procedures, Workflow, and Integration	Expert Level with Understanding the Applications of GIS Software	Expert with GIS Training, Education, and Knowledge Transfer	Understands IT Infrastructure and Architecture
DESIGN AND WORK WITH FUNCTIONAL GIS GROUPS Work well with GIS functional groups.	**MANAGE AND MAINTAIN CRITICAL DATA LAYERS** The GIS Coordinator should pay particular attention to the base layers of the enterprise GIS including: • Parcels • Address Points • Street Centerlines • Aerial Photography	**DEVELOP DATA MAINTENANCE PROCEDURES** Data maintenance procedures. A critical SOP will be the procedures and protocols for maintaining digital data layers.	**PLAN, DESIGN, AND DEPLOY PUBLIC ACCESS SOLUTIONS** The GIS Coordinator is ultimately responsible for planning and designing an effective public access portal.	**CONDUCT ROI- VALUE PROPOSITION, AND COST BENEFIT ANALYSIS EDUCATION WORKSHOPS** The GIS Coordinator is ultimately responsible for conducting ROI workshops.	**TRAIN AND EDUCATE IT STAFF** The GIS Coordinator is ultimately responsible for conducting GIS training for IT professionals.
CONDUCT GIS USER GROUP MEETINGS. Lead GIS user group meetings. Ultimately in charge. **MANAGE THE REGIONALIZATION OF GIS** Understand the importance of Regionalization of GIS (and can develop MOU and data sharing agreements)	**DEVELOP AN ENTERPRISE DATABASE DESIGN** Develop an Enterprise database design. The GIS Coordinator should have the ability to plan, design, and implement an enterprise database.	**CREATE APPLICATION DEVELOPMENT PROCEDURES** The GIS Coordinator will be ultimately responsible for developing GIS "application and acquisition/ development" procedures.	**DEPLOY AND MANAGE WEB BASED GIS SOLUTIONS** The GIS Coordinator is ultimately responsible for deploying an online iritiative.	**CONDUCT KNOWLEDGE TRANSFER THROUGHOUT THE ORGANIZATION** The GIS Coordinator is ultimately responsible for encouraging and designing knowledge transfer sohutions, including GIS day, Steering Committee meetings, and user group discussions.	**SECURE 24/7 AVAILABILITY OF IT GIS INFRASTRUCTURE** The GIS Coordinator is ultimately responsible for supporting 24/7 availability.
CREATE AND RATIFY GIS POLICY Develop GIS policy and mandates. Developing and enforcing Standard Operating Procedures and policies related to GIS.	**CREATE DATA CREATION POLICY** Develop data creation policies and procedures. The GIS Coordinator should create and enforce data creation policies and procedures.	**DEFINE METADATA** Define Metadata standards. The GIS Coordinator will need to define Metadata standards.	**DEPLOY AND MAINTAIN GIS CENTRIC STORY MAPS** The GIS Coordinator is ultimately responsible for using simple and effective GIS communication tools.	**MANAGE EDUCATIONAL CONFERENCE ATTENDANCE** The GIS Coordinator is ultimately responsible for attending conferences and organizing staff to attend conferences.	**SECURE ENTERPRISE BACK-UP PROCEDURES AND PROTOCOLS** The GIS Coordinator is ultimately responsible for supporting enterprise back-up.
MONITOR GIS USAGE AND STAKEHOLDER NEEDS Sensitive to stakeholders. Departmental stakeholder (client) support requires feedback through meeting and questionnaire.	**MANAGE THE CENTRAL GIS REPOSITORY** Create and maintain a central GIS repository. The GIS Coordinator should be responsible for one central repository rather than silos of data.	**PREVENT DATA DUPLICATION** Prevent data duplication between systems. The GIS Coordinator should be fully aware of any duplication that exists within the organization and identify remedies and solutions.	**DEPLOY AND MANAGE CROWDSOURCING SOLUTIONS** The GIS Coordinator is ultimately responsible for implementing crowdsourcing applications.	**PROMOTE ONLINE GIS TRAINING / EDUCATIONAL SEMINARS AND WORKSHOPS** The GIS Coordinator is ultimately responsible for administering online seminars and workshops.	**UNDERSTAND DATA STORAGE REQUIREMENTS PROCEDURES AND PROTOCOLS** The GIS Coordinator s ultimately responsible for understanding data storage procedures and protocols.

FIGURE 7.11 (CONTINUED)
Characteristics of a perfect GIS coordinator.

(Continued)

Characteristics of an Enterprise GIS Coordinator or Geographic Information Officer (GIO)

Great Communicator and Presenter • Educator • Problem Solving and Analytical Skills • Integrity • Competence • Ability to Delegate
Empathy • Understand the Art of Cooperation • Enthusiasm and Passion • Inspires a Shared Vision, Goals, and Objectives • Team-Building Skills

GIS Governance and Management Skills	GIS Digital Data and Databases Expertise	Understand and Documents Procedures, Workflow, and Integration	Expert Level with Understanding the Applications of GIS Software	Expert with GIS Training, Education, and Knowledge Transfer	Understands IT Infrastructure and Architecture
HOW TO COLLABORATE WITH ALL STAKEHOLDERS Understands GIS collaboration. GIS collaboration is the hallmark of an enterprise GIS.	**PLAN, DESIGN, AND DEPLOY MOBILE GIS** Implement mobile solutions. The GIS Coordinator should be ultimately responsible for the mobility of GIS including standards, procedures, hardware, operating systems, and communication.	**ENTERPRISE INTEGRATION AND INTEROPERABILITY** The GIS Coordinator should understand and implement a high level of integration and interoperability between systems.	**ENABLE ELECTED OFFICIALS AND DECISION MAKERS WITH EASY TO USE GIS SOLUTIONS** The GIS Coordinator is ultimately responsible for enabling elected Officials with GIS.	**MANAGE KNOWLEDGE TRANSFER GIS BROWN BAG LUNCH MEETINGS** The GIS Coordinator is ultimately responsible for conducting brown bag lunches.	**TAKE RESPONSIBILITY FOR HARDWARE NEEDS OF THE ENTERPRISE GIS** The GIS Coordinator is ultimately responsible for understanding IT, hardware, and mobile standards.
MEASURE SERVICE LEVELS OF THE GIS PROGRAM Implement and measure quality of GIS service. **UNDERSTAND ORGANIZATIONAL THEORY AND PRACTICE** Understands GIS authority and clear lines of responsibility.	**CREATE AN OPEN DATA OPEN GOVERNMENT STRATEGY** Develop Open data open government initiative. The GIS Coordinator is ultimately responsible for sharing accurate and reliable information.	**IDENTIFY THE KEY GIS INTEGRATION POINTS IN LOCAL GOVERNMENT** It is the GIS Coordinator's role to integrate GIS with the following: • Work order solutions • ERP solutions (permitting) • Public safety	**UTILIZE GIS EXTENSIONS TO IMPROVE DEPARTMENTAL USE OF GIS** The GIS Coordinator is ultimately responsible for understanding and deploying modeling extensions.	**OVERSEE GIS SUCCESSION PLANNING** The GIS Coordinator is ultimately responsible for understanding succession planning.	**DEVELOP A MOBILE ACTION PLAN** The GIS Coordinator is ultimately responsible for developing a GIS mobile action plan.
UNDERSTAND GIS FUNDING MODELS, BUDGETING AND ACCOUNTING Manage a GIS budget through a funding model **WRITE GRANT APPLICATIONS OR PROPOSALS** Develop Grants and funding initiatives.	**DEVELOP QUALITY ASSURANCE / QUALITY CONTROL PROCEDURES** The GIS Coordinator should enforce QA/QC procedures on all digital GIS data layers.	**MANAGE GIS TECHNICAL SUPPORT** The GIS Coordinator should implement and enforce GIS technical support (ticketing/ help desk).	**DEPLOY MOBILE GIS SOLUTIONS** The GIS Coordinator is ultimately responsible for deploying mobile GIS software.	**CONDUCT CREATIVE AND INNOVATIVE THINKING IN LOCAL GOVERNMENT** The GIS Coordinator will be required to encourage and promote innovative thinking through the use of Blue Sky Sessions.	**DEPLOY GIS STAGING AREA** The GIS Coordinator is ultimately responsible for supporting and understanding a GIS staging and development zone.
DEVELOP DETAILED ANNUAL WORK PLAN AND SERVICE LEVEL AGREEMENTS AND MOUs: Create an annual detailed GIS work plan. **DEVELOP KEY PERFORMANCE MEASURES FOR THE GIS PROGRAM** Create Key Performance measures or Indicators (KPIs).		**DOCUMENT AND DETAIL ALL GIS WORKFLOWS** Extensive workflow documentation of GIS editing, data management, data analysis, data publication, map production, and enterprise GIS management.	**MANAGE THE USE OF GPS TECHNOLOGY THROUGHOUT THE ORGANIZATION** The GIS Coordinator is ultimately responsible for using Global Positioning System (GPS) technology.	**CONDUCT TEAM BUILDING SKILLS** The GIS Coordinator should encourage the use of GIS technology throughout the organization.	**PERFORM SYSTEM ADMINISTRATION DUTIES** The GIS Coordinator should understand the basics of systems administration and the benefits to an enterprise GIS.

FIGURE 7.11 (CONTINUED)
Characteristics of a perfect GIS coordinator.

Characteristics of an Enterprise GIS Coordinator or Geographic Information Officer (GIO)

Great Communicator and Presenter • Educator • Problem Solving and Analytical Skills • Integrity • Competence • Ability to Delegate • Empathy • Understand the Art of Cooperation • Enthusiasm and Passion • Inspires a Shared Vision, Goals, and Objectives • Team-Building Skills

GIS Governance and Management Skills	GIS Digital Data and Databases Expertise	Expert Level with Understanding the Applications of GIS Software	Understand and Documents Procedures, Workflow, and Integration	Expert with GIS Training, Education, and Knowledge Transfer	Understands IT Infrastructure and Architecture
USE SOCIAL MEDIA TO PROMOTE GIS SUCCESS Develop a GIS newsletter. CULTIVATE A GIS CULTURE Create GIS culture of collaboration.				DEVELOP A TRAINING, EDUCATION AND KNOWLEDGE TRANSFER STRATEGY The GIS coordinator is ultimately responsible for developing and branding an organization-wide solution	MANAGE AND TAKE RESPONSIBILITY FOR DATA SECURITY: The GIS Coordinator should take responsibility for data security of the enterprise GIS.
UNDERSTAND THE IMPORTANCE AND BENEFIT OF ALIGNMENT Align GIS with organization's vision, goals, and objectives.					
SELL GIS TO THE LOCAL GOVERNMENT ORGANIZATION The GIS Coordinator needs to be able to sell GIS to the organization.					
OVERSEE AND MANAGE ALL NEW AND INNOVATIVE GEOGRAPHIC TECHNOLOGIES The GIS Coordinator needs to be able to understand new solutions and local government application. An example would be the use of DRONES.					

FIGURE 7.11 (CONTINUED)
Characteristics of a perfect GIS coordinator.

- Mentally demanding and hard to sustain.
- Do things that are not inherently enjoyable.

I used Colvin's national bestseller book *Talent Is Overrated* to try to formulate a plan of action for improving the performance of all GIS professionals and, especially, the person who is ultimately in charge: the GIS coordinator or the GIO. So, let us think about this for a moment. We need training, education, and knowledge transfer tools that address Colvin's four principles of deliberate practice. They do not exist in the GIS world. We still focus primarily on how to use software with a sprinkling of theory and management workshops. What about a workshop that is engineered to train, educate, and transfer knowledge that embraces Colvin's principles? An example of a workshop is as follows:

Three-day GIS workshop: *The Value of GIS in Local Government*

- The story (design practice)
- Presentation (high repetition)
- Evaluation (feedback)
- Facts (mentally demanding)
- Practice (inherently not enjoyable)

7.6.2 Idea Number Two: Modern-Day Skills of a GIS Coordinator

It is best to refer to Colvin's work again when it comes to describing the characteristics of a new generation of GIOs. According to a summary of Colvin's book, *Humans Are Underrated*, we must remember that in the near future, "Our greatest advantage lies in what we humans are most powerfully driven to do for and with one another, arising from our deepest and most essentially human ability, empathy, creativity, social sensibility, storytelling, humor, building relationships, and expressing ourselves with greater power than logic can ever achieve." This approach will potentially offer the GIS world a more devoted GIS user base, a stronger GIS culture, in addition to new and creative ideas. Colvin points out that even as technology advances, the richness of interpersonal experience will continue to define us in the future.

After 20 years of consulting, I can confirm the fact that we have more success working with GIS professionals who are unafraid to ask for advice or directions. More often than not, it is women who ask for directions. According to both my experience and the ideas of Colvin, women are more empathetic; they understand the thoughts and intentions of others, understand nonverbal communication and value reciprocal relationships, are cooperative and collaborative, and understand fairness more than their male counterpart. So, I just want to point out to the readers, by way of Colvin, something that is near-universally accepted though often with tongue in cheek; women ask for directions, whereas men do not. The point here is that the research indicates that women are potentially better prepared for an advanced technological future than men. There is hope for men though; however, they need to embrace Colvin's key human traits that will enable them to be more prepared for the future. And these traits are listed as follows:

- Empathy
- Cooperation
- Collaboration
- Relationships
- Equal conversation
- Communication
- Social abilities

My recommendation is to integrate this mind-set into a new, improved strategy for GIS training, education, and knowledge transfer within local government.

Three-day GIS workshop: *GIS Collaboration and Regionalization*

- Introduction to GIS collaboration and regionalization
- Communication
- Cooperation
- Collaboration
- Empathy
- Relationships
- Formalizing a relationship
- Best business practices in collaboration and regionalization

7.6.3 Idea Number Three: The Principles of the CCL

The CCL supports, and I think formalizes, some of Colvin's ideas. The CCL collaborates with many organizations, such as Ravenscroft School in Raleigh, North Carolina. Ravenscroft School represents a new approach to a rapidly advancing technological world. Doreen Kelly, Head of Ravenscroft School, was quoted, "Game-changing graduates are bold and courageous, aware, compassionate, able and successful... and prepared to be leaders for a radically changing and extraordinarily challenging world."

From that statement, we could gather that the goal of Ravenscroft School is to intentionally weave leadership and citizenship learning into the fabric of the school and community. Ravenscroft is an independent pre K–12 school that is celebrating its one-hundred fiftieth anniversary this year, and its legacy runs deep within the extended community of students, alumni, faculty, staff, and families. The school makes its focus on the future possible by weaving the following principles into the curriculum:

- Leadership
- Citizen development
- Entire culture development
- Common understanding
- Belief
- Behavior
- Leadership
- Experiential learning
- Sense of community
- Resilient
- Capable

Here is another quote from the school that, with a few tweaks, could fit snugly into the context of geospatial technologies:

> Ongoing efforts are geared to the integration of the framework and practices throughout the school, including weekly consultations with the CCL youth leadership team. By 2015, every student and teacher will be fully engaged, along with Ravenscroft's executive leadership, board, parent body and alumni. Our commitment is to developing young people as leaders and citizens. We have begun to change the education conversation away from quick fixes and the latest 'reforms' and toward preparing students to be resilient, capable and able to lead in a complex and interdependent world.

My recommendation is to integrate CCL's leadership strategy into GIS training, education, and knowledge transfer within local government. An example would be as follows:

Three-day GIS workshop: *GIS Leadership*

- Leadership: What is it?
- Developing a GIS culture.
- Professional GIS behavior and ethics.
- Collaboration and understanding.
- Empathy and relationships.
- Creative thinking.
- Belief and resilience.

We now have three ideas that can influence modern-day training, education, and knowledge transfer solutions. In my humble opinion, these are three things that are simultaneously missing from the world of GIS and local government professionals and are important to the future of the industry.

To translate the prior three components into reality, I created a checklist of the characteristics of a perfect GIS coordinator. The idea is that we weave the three principles of education into the requirements of a future GIS coordinator and GIS staff.

The following represent my ideas of a comprehensive training plan (it not a GIS curriculum) that not only embraces the current GIS needs in local government but will also support a new and innovative future approach.

7.7 GIS Training Series Module One: Characteristics of the Project GIS Coordinator

The following documents the characteristics of a future GIS coordinator and the required training:

• Great communicator	Communication skills
• Expert presenter	Presentation skills
• Educator	Teaching skills
• Problem-solving and analytical skills	Management skills
• Integrity	Ethical skills
• Competence	Management skills
• Ability to delegate	Management skills
• Empathy	Management skills
• Understand the art of cooperation	Management skills
• Enthusiasm and passion	Management skills
• Inspires a shared vision, goals, and objectives	Management skills
• Team-building skills	Management skills

The following list was developed for a local government to fill a gap in its training, education, and knowledge transfer requirements.

Understanding GIS and Local Government Series
GIS Workshops

Introduction Series

1. *Introduction to Geographic Information Systems (GIS)*: What is GIS?

2. *GIS Strategic Planning and Management*: Planning, designing, and managing an enterprise and sustainable GIS

Software Series

1. *Exploring the World of Esri Software*: Demonstrating the Esri software suite

2. *Local Government Business Applications*: How to effectively use GIS in local government operations

3. *The Mobility of GIS*: How mobile is GIS? Using GIS in the field to collect, maintain, and update your databases

4. *GIS and Public Safety Operations*: Utilizing GIS for public safety and emergency operations

5. *GIS and the Natural Environment*: Using GIS to manage parks, land, and natural resources

Management Series

1. *GIS: A Profitable Initiative*: Creating a business case for GIS

2. *The Architecture of GIS*: Critical GIS components

3. *The GIS Obstacle Course*: Overcoming GIS challenges, barriers, and pitfalls

4. *Enterprise GIS: State of the Union*: An organization's GIS status

5. *The Future of GIS*: The future of GIS technology

7.8 GIS Training Series Module Two: GIS Governance and Management Skills

The following is a list of training modules that are required in local government.

- *Strategic planning*—Understand the GIS strategic planning process and the benefits of a detailed plan.
  - Conducting an annual strategic assessment
- *Developing a vision, goals and objectives*—Develop a GIS vision, goals, and objectives for the organization. Tools include Blue Sky sessions, online questionnaires, and departmental interviews.
- *Understand governance model*—Enforce and manage a formalized governance model. Secure top-down support for a ratified enterprise governance model.

- *Developing detailed GIS job classifications and duties*—Understand job duties and job classifications. Collaborate with subject matter experts (SMEs) on GIS projects and initiatives.
- *Enterprise project management*—Manage and monitor all departmental projects with emphasis on the enterprise aspects.
- *A coordinated enterprise GIS*—Coordinate all GIS activities, projects, and protocols. Synchronize cross-departmental intentions and communication.
- *Effective communication across all departments and the enterprise*—Communicate effectively to the GIS steering committee. Regularly and effectively present the state of the GIS technology using graphic-rich media.
- *How to design and work with functional GIS groups*—Work well with GIS functional groups. Communicate and collaborate with all GIS teams, functional groups, and departmental SMEs.
- *How to conduct GIS user group meetings*—Improve the quality, content, and schedule of the GIS user group meetings.
- *How to manage the regionalization of GIS*—Understand the importance of the regionalization of GIS, and develop the subsequent MoU and data-sharing agreements.
- *How to create and ratify GIS policy*—Develop and enforce SOPs that are related to the GIS.
- *How to monitor GIS usage and stakeholder needs*—Support departmental stakeholders via the feedback that is gained through meetings and questionnaires.
- *How to collaborate with all stakeholders*—Understand the collective processes that are a hallmark of an enterprise GIS.
- *Measure service levels of the GIS program*—Gauge the quality of GIS services by using questionnaires and annual surveys. Additionally, a call record for GIS service is critical to measure GIS support.
- *Understand organizational theory and practice*—Understand authority and clear lines of responsibility. The ability to diagram lines of authority and responsibilities within the organization is critical.
- *Understand GIS funding models, budgeting, and accounting*—Manage a GIS budget through a funding model.
- *How to write grant applications or proposals*—Develop grants and funding initiatives. Grants are important to a successful GIS, and thus working with the grant writer is an important component of the process.
- *How to develop a detailed annual work plan and service level agreements and MOUs*—The GIS coordinator should be ultimately in charge of

an annual GIS action plan that supports the enterprise. This should be shared with all departments.

- *Developing key performance measures for the GIS program*—Using the vision, goals, and objectives, the GIS coordinator should develop key performance measures that quantify success.
- *How to use social media to promote GIS success*—Develop a GIS blog or newsletter to promote the success of the technology. Do not underestimate the importance of this aspect of enterprise GIS.
- *Cultivating a GIS culture within the organization*—The GIS coordinator should cultivate an environment of collaboration and develop a community that speaks the language of GIS.
- *Understanding the importance and benefit of alignment*—The GIS coordinator should be ultimately responsible for aligning the GIS vision with the organization's vision.
- *How to sell GIS to the local government organization*—The GIS coordinator must be able to sell the GIS to the organization.

7.9 GIS Training Series Module Three: GIS Digital Data and Databases Expertise

- *How to perform a digital data assessment*—The GIS coordinator should be able to assess the quality, content, and completeness of every digital data layer.
- *Creating a comprehensive master data list with custodianship*—The GIS coordinator should maintain an accurate and reliable master data list and share this with all departments.
- *Developing metadata standards and how to enforce the creation of metadata*—The creation of metadata for every digital data layer should be encouraged and enforced by the GIS coordinator.
- *How to oversee data layer maintenance*—The GIS coordinator should ensure that critical data layers are kept up to date. Every digital data layer should have a departmental custodian who maintains the accuracy, content, and completeness of that digital data layer.
- *How to manage and maintain critical data layers*—The GIS coordinator should pay particular attention to the base layers of the enterprise GIS including the following:
  - Parcels
  - Address points

- Street centerlines
- Aerial photography

- *How to develop an enterprise database design*—The GIS coordinator should have the ability to plan, design, and implement an enterprise database.

- *How to create a data creation policy*—The GIS coordinator should create and enforce data creation policies and procedures.

- *How to manage the central GIS repository*—The GIS coordinator should be responsible for creating and maintaining a single central repository, rather than silos of data.

- *Planning, designing, and deploying mobile GIS*—The GIS coordinator should be ultimately responsible for implementing mobile GIS solutions including standards, procedures, hardware, operating systems, and communication.

- *Create an open data/open government strategy*—Develop an open data/open government initiative. The GIS coordinator is responsible for sharing accurate and reliable information.

- *Develop quality assurance/quality control procedures*—The GIS coordinator should enforce quality assurance/quality control procedures on all digital GIS data layers.

7.10 GIS Training Series Module Four: Understand Procedures, Workflow, and Integration

- *How to oversee enterprise GIS integration*—The GIS coordinator should understand the business and RoI opportunities for integrating the GIS with existing legacy solutions.

- *Techniques to identify gaps and opportunities*—All the gaps and opportunities for GIS integration should be identified and presented by the GIS coordinator.

- *Document and detail creative and innovative departmental GIS use*—Create departmental solutions for access to critical data layers. The GIS coordinator should have the skill set to plan, design, and implement departmental GIS solutions.

As we speak, I am in Alaska using a drone to capture images that are critical to decision support. Figure 7.4 shows a colleague using a drone to gather a video of a development site on the Aleutian Islands.

- *Developing SOPs*—Develop GIS SOPs. The GIS coordinator is ultimately responsible for developing enterprise and departmental SOPs.
- *Developing data maintenance procedures*—Data maintenance procedures. A critical SOP will be the procedures and protocols for maintaining digital data layers.
- *Creating application development procedures*—The GIS coordinator will be ultimately responsible for developing GIS *application and acquisition/development* procedures.
- *Defining metadata*—The GIS coordinator needs to define metadata standards.
- *How to prevent data duplication*—Prevent data duplication between systems. The GIS coordinator should be fully aware of any duplication that exists within the organization and identify remedies and solutions.
- *Evaluating enterprise integration and interoperability*—The GIS coordinator should understand and implement a high level of integration and interoperability between systems.
- *Identifying the key GIS integration points in local government*—It is the GIS coordinator's role to integrate the GIS with the following:
  - Work order solutions
  - Enterprise resource planning solutions
  - Public safety
- *How to manage GIS technical support services*—The GIS coordinator should implement and enforce GIS technical support and be sure that a ticketing and help desk system is incorporated.
- *How to document and detail all GIS workflows*—Extensively document the workflow data regarding GIS editing, data management, data analysis, publication, map production, and enterprise GIS management.

7.11 GIS Training Series Module Five: Understanding the Applications of GIS Software

- *How to manage GIS software licenses*—The GIS coordinator is ultimately responsible for managing GIS licenses and any enterprise license agreements.
- *Understanding the issues of commercial off-the-shelf versus custom code*—The GIS coordinator is responsible for enforcing policies about the level of GIS commercial off-the-shelf versus custom code.

- *Enabling enterprise GIS accessibility*—The GIS coordinator is responsible for continually improving access to software.
- *How to plan, design, and deploy Intranet GIS solutions*—The GIS coordinator is ultimately responsible for planning, designing, and implementing an Intranet solution.
- *How to plan, design, and deploy public access solutions*—The GIS coordinator is ultimately responsible for planning and designing an effective public access portal.
- *How to deploy and manage Web-based GIS solutions*—The GIS coordinator is ultimately responsible for deploying an online initiative.
- *How to deploy and maintain GIS-centric story maps*—The GIS coordinator is ultimately responsible for using simple and effective GIS communication tools.
- *How to deploy and manage crowdsourcing solutions*—The GIS coordinator is ultimately responsible for implementing crowdsourcing applications.
- *How to enable elected officials and decision makers with easy-to-use GIS solutions*—The GIS coordinator is ultimately responsible for empowering elected officials with the GIS.
- *Utilizing GIS extensions to improve departmental use of GIS*—The GIS coordinator is ultimately responsible for understanding and deploying modeling extensions.
- *Deploying mobile GIS solutions*—The GIS coordinator is ultimately responsible for deploying mobile GIS software.
- *How to manage the use of GPS technology throughout the organization*—The GIS coordinator is ultimately responsible for using GPS technology.

7.12 GIS Training Series Module Six: GIS Training, Education, and Knowledge Transfer

- *Developing a formal GIS training plan*—The GIS coordinator is ultimately responsible for creating a formal ongoing GIS training plan.
- *Conducting multitiered GIS training and education*—The GIS coordinator is ultimately responsible for understanding multitiered GIS software training.
- *Conducting mobile GIS training*—The GIS coordinator is ultimately responsible for administering mobile software training.

- *How to conduct departmental GIS training*—The GIS coordinator is ultimately responsible for administering departmental specific education.
- *How to conduct RoI/value proposition/cost–benefit analysis education workshops*—The GIS coordinator is ultimately responsible for conducting RoI workshops and regularly evaluating the value of GIS to the organization.
- *Conducting knowledge transfer throughout the organization*—The GIS coordinator is ultimately responsible for encouraging and designing knowledge transfer solutions, including GIS Day, steering committee meetings, and user group discussions.
- *Managing educational conference attendance*—The GIS coordinator is ultimately responsible for attending conferences and organizing staff to attend conferences.
- *Promoting online GIS training/educational seminars and workshops*—The GIS coordinator is ultimately responsible for administering online seminars and workshops.
- *How to manage knowledge transfer GIS brown bag meetings*—The GIS coordinator is ultimately responsible for conducting brown bag meetings.
- *How to oversee GIS succession planning*—The GIS coordinator is ultimately responsible for understanding succession planning, and enforcing its use.
- *How to promote creative and innovative thinking in local government*—The GIS coordinator will be required to encourage and promote innovative thinking through the use of Blue Sky sessions and more.
- *How to conduct team-building skills*—The GIS coordinator should encourage the use of GIS technology throughout the organization.

7.13 GIS Training Series Module Seven: Understands IT Infrastructure and Architecture

- *Understanding strategic IT planning for an enterprise GIS*—The GIS coordinator is ultimately responsible for understanding the strategic technology plan. It is critical to understand IT technology plans as it relates to the GIS initiative.
- *How to create an architectural design for an enterprise GIS*—The GIS coordinator is ultimately responsible for developing a high-level GIS architectural design and presenting his solution to decision makers.

- *How to monitor infrastructure requirements of an enterprise GIS*—The GIS coordinator is ultimately responsible for documenting and detailing IT infrastructure to meet operational needs.
- *How to understand IT replacement planning*—The GIS coordinator is ultimately responsible for understanding the IT replacement plan especially as it relates to GIS.
- *How to train and educate IT staff*—The GIS coordinator is ultimately responsible for conducting GIS training for IT professionals.
- *Understanding secure 24/7 availability of IT–GIS infrastructure*—The GIS coordinator is ultimately responsible for supporting 24/7 availability.
- *Understanding secure enterprise backup procedures and protocols*—The GIS coordinator is ultimately responsible for supporting enterprise backup of all GIS data and data layers.
- *Understanding data storage requirements and procedures and protocols*—The GIS coordinator is ultimately responsible for understanding data storage procedures and protocols.
- *How to take responsibility for hardware needs of the enterprise GIS*—The GIS coordinator is ultimately responsible for understanding IT, hardware, and mobile standards.
- *Developing a mobile action plan*—The GIS coordinator is ultimately responsible for developing a GIS mobile action plan.
- *Understanding how to deploy a GIS staging area*—The GIS coordinator is ultimately responsible for supporting and understanding a GIS staging and development zone.
- *How to perform system administration duties*—The GIS coordinator should understand the basics of systems administration and the benefits to an enterprise GIS.
- *How to manage and take responsibility for data security*—The GIS coordinator should take responsibility for the data security of the enterprise GIS.

8

Return on Investment

It's not hard to make decisions once you know what your values are.

Roy E. Disney

8.1 Introduction

Your greatest weakness as a GIS coordinator is failing to measure the value of GIS. I have literally traveled to every corner of the United States and worked with hundreds of local government organizations. Without exception, no organization has ever attempted to measure or even promote the real tangible and intangible benefits or value of GIS. OK, there may be a couple of them.

In this chapter, I introduce 16 return on investment (RoI) categories that are used to explain, document, and detail real-world examples of how GIS benefits government organizations. This section discusses how GIS technology can save lives, inform and notify the public, prevent local governments from being fined, improve the efficiency of virtually all departments, eliminate duplication, and predict events and infrastructure failures, in addition to improving management in all areas of the organization.

So, the question in this chapter is, is quantifying the benefits of GIS purely a mathematics problem, or is it a problem of perception?

My goal in this chapter is to explore this question. It is not to introduce a calculator that *spits out* an RoI value that is measured in dollars or offer new and innovative ways to perform a cost–benefit analysis (CBA). Rather, I will lay out the differences and similarities between some of the most common buzzwords in the industry, including a CBA, a RoI assessment, and a value proposition (VP).

I am hoping that after reading this chapter, you will be able to confidently present to your elected officials why they cannot live without this all-encompassing, unique management tool that brings incredible benefit to the organization and the community.

Strategic GIS planning raises a threefold question that focuses on implementation, contribution, and benefit. Firstly, how does a town, city, or county integrate GIS into their existing work practices and corporate culture? Secondly, how does the GIS contribute to functions and operations? And thirdly, what is the value of this technology to the organization, and why should any organization invest, or continue to invest, in this technology? It is the third question that we are going to wrestle with in this chapter.

8.2 Local Governments' Scorecard Approach

Over the past decade, many towns, cities, and county governments have embraced a *scorecard* approach to measuring operations and services. The following are examples of local government use of the scorecard approach.

8.2.1 Example One: The City of St. Petersburg, Florida

The City of St. Petersburg in Florida has a very interesting city scorecard that is *composed of performance measures for various departments and services*. As the city indicates on its Website, "These performance measures allow members of the community to follow the progress of the city and keep informed as to how their city government is performing." The City of St. Petersburg gauges performance metrics in the following categories:

- Website pages viewed
- Number of permits issued
- Available real estate
- Number of citizen complaints
- Number of arrests
- Number of emergency calls
- Crime rate by type of crime
- Traffic fatalities

8.2.2 Example Two: The City of Boston, Massachusetts

There are a variety of slightly different goal measuring strategies that are available to local government. The City of Boston provides a pertinent example in its deployment of a program called *Boston About Results*. It is obviously about setting the *bar* high enough. The City of Boston explains on its Website that their metrical program covers the "city's performance management program that uses performance measurement and data analytics to develop strategies and programs that evaluate city performance, reduce costs, and ultimately deliver better services to Boston's residents, businesses, and visitors." This 2016 Website includes statistics or powerful statements about many services, again, including, but not limited to, the following meters of effectiveness:

- Library use increased by 20%.
- The health division inspected over 1400 restaurants.
- Police have removed 515 illegal crime guns.
- The parks department completed an all-time high of 1632 maintenance requests.
- Job creation increased by 5%.
- Two hundred graffiti removed.
- Eighty-one fewer fires in quarter 1.

Boston's performance measurement program is not wholly limited to quantifiable positive or negative statistics. Most statements are facts about performance, rather than a scorecard approach that assigns passing or failing grades. They are powerful nonetheless.

8.2.3 Example Three: The City of Fort Collins, Florida

The City of Fort Collins has a community scorecard that is a collection of performance data and information that measures how well the city is doing in meeting its goals in the following seven areas:

1. Community and neighborhood livability
2. Culture and recreation
3. Economic health
4. Environmental health
5. High-performing government
6. Safe community
7. Transportation

The City of Fort Collins 2011 Community Scorecard is a brightly colored and attractive online document presenting the statistics, in addition to details about departmental and community awards. The city's data come from operational information that is supplied by each city department and the International City/County Management Association Center for Performance Measurement. The city supports the statistics with some visually pleasing line graphs, bar charts, and pie charts, which are a transparent and accessible means for everyone to access this information.

It would seem that there is enough evidence to support the notion that local government organizations are no stranger to performance metrics and a scorecard approach to their community performance. So the question remains, why hasn't there been more effort to evaluate the benefits of GIS? Let us look at some industry-standard options for measuring the benefits of GIS.

8.3 Option One: CBA

A *CBA* is used to analyze the balance between the costs and benefits of new technology. It includes evaluating all of the costs and all of the benefits, as well as the risk and reward of maintaining the status quo or the *do-nothing* option. It is, as the name implies, a simple comparison of the cost of everything to do with GIS implementation versus the quantifiable benefits of implementation.

A study of this nature evaluates and ranks alternative solutions based on how much the quantifiable benefits outweigh the cost. Simply put: a CBA is a systematic approach to quantifying the economic value of a solution that is measured in dollar value for the cost of implementation versus the benefits to the organization. In local government, this type of study can range from a 300-page comprehensive analysis to a one-page, back-of-the-envelope summary.

A CBA becomes complicated when attempts are made to quantify the units of measurement that are, by their nature, based in subjectivity. These stats come across in categories like satisfaction of users, health and wellbeing, safety and security, and perceived quality of human life. Though some say that there are monetarily unquantifiable benefits of GIS, it is important to remember that this type of study is no more than a well-educated estimate and tends to focus on the following:

- Accurately calculating costs.
- Calculating the benefits that are measured in dollars.
- The difference between costs and benefits equals value.
- A scientific approach but still a well-educated estimate.
- It does measure risk and indirect and intangible benefits.

8.4 Option Two: RoI Analysis

An *RoI* analysis is less comprehensive than a CBA and is focused exclusively on the benefits of an investment. It is more about establishing metrics or performance measures and calculating the *rate of return*. A high RoI means that the benefits significantly outweigh the costs. RoI metric categories are essentially economic performance measures used to determine the effectiveness of an investment. The purpose of an RoI is to measure the rate of return on the money that is invested in the solution. The RoI, however, may be calculated in terms other than financial metrics. It can be used by any entity to evaluate impact on stakeholders, identify ways to improve performance, and enhance the performance of investments. It is important to know the present state or existing conditions of an organization in order to develop appropriate performance measures for an RoI. This type of analysis focuses on the following:

- Selecting and establishing performance metrics are determined by what is important to that organization or department.
- Understanding the present state of the organization.
- Focusing on the benefits of the metrics.
- Emphasizing the rate of return.
- Unconcerned with intangibles.

8.5 Option Three: A VP

A *value proposition* is the least intimidating analysis option because it is based on the premise of a *promise of value*. This refers to a belief on behalf of elected officials and decision makers, taken on faith in the technology, that the value of GIS will be delivered and experienced without any presentation of empirical evidence. For your information, I like this analysis the most, as it gives us plenty of room for emotion.

Like the CBA, a VP is based on reviewing and analyzing the costs and benefits. The difference is that if the perceived benefits theoretically outweigh the perceived costs, the solution is deemed worthy of implementation. A VP is inherently less scientific and is concerned with the perception of GIS in local government, as *opposed to the quantifiable realities* of GIS in local government.

The VP is the foundation of an elevator pitch, which, as the name implies, is a simple, understandable, and quick sales pitch that could be delivered in a two-minute elevator ride. More importantly, a VP puts more weight on the fact that the value or benefit of GIS is perceived differently by different people—that is to say that stakeholders within local governments will weight benefits differently.

A VP is a clear, simple, and short statement of the benefits, both tangible and intangible, that the GIS will provide to the local government organization, along with the price of implementation. VPs generally focus on the following:

- Review and analysis of costs.
- Promise of a value return and the detailed perceived benefits.
- The weighted perceived benefits should outweigh the costs.
- A humanistic sales and elevator pitch approach.
- Limited on empirical quantification.

VPs, along with CBA and RoI analyses, allow an organization to examine an investment and make good decisions. In summation, the RoI is valuable when costs and benefits are tangible, whereas the CBA is more useful when both the tangible and intangible costs and benefits are considered. A VP takes a holistic view of the benefits and promises value based on good judgment.

The level of effort put into one of the three strategies for assessing the value of an investment is a critical decision.

8.6 Perceived Benefits

Most government organizations do not have a formal framework or standardized payback model for measuring how much value GIS delivers to their organization. The question is, does this matter?

In the modern world, local government organizations are required to do more with less. Since 2008, it would seem that every town, city, and county in the United States has had to tighten their belts when it comes to technological investment. Some academics insist that local governments will refuse to implement GIS technology without demonstrable, quantified benefits. After twenty-five years roaming the countryside and developing successful roadmaps for local government, I cannot agree with this statement.

I concede that providing a compelling explanation of GIS technology and its potential benefit to an organization is essential, and I also agree that local government should use a formula to quantify benefits. That all being said, I believe that the success of GIS implementation has more to do with the way the implementation is perceived by decision makers. For that to happen, the value of GIS must be demonstrated and, to a certain extent, quantified.

In my experience, no local government organization has ever refused to implement GIS technology because they were not 100% certain that potential benefits would eventually outweigh the costs of implementation. Risks and rewards come with every project, and local governments understand this. Accordingly, I advise against any preoccupation with numbers and suggest a concerted focus on crafting a compelling argument regarding the benefits of GIS to be delivered to the decision-making individuals in local government.

8.7 More Trouble Than It's Worth

It is important to know all the barriers, pitfalls, and problems that are associated with any plan before embarking on a project. This holds true in all aspects of life. Thus, we need to acknowledge the following points:

- Developing a value for business outcomes and operational performance improvements is very difficult to do.
- Some projects are unsuitable for RoI analysis.
- Even after performing a rigorous RoI analysis, local government organizations may not achieve the results that are promised.
- Setting expectations is a key factor for success.
- Many RoI projects have a flawed approach to quantifying the benefits.
- Measuring changed processes and people costs is very difficult, and quantifying how long people take to perform a task is hard.
- An RoI analysis is not a science. It is impossible to calculate accurately.
- It is most successful and suitable for the most basic, straightforward projects.
- A formal measurement framework means creating a consistent way of evaluating the current and future information technology projects in terms of how they affect the business.
- You require an understanding of what your business perceives to be of value.

How can we possibly quantify the ways that e-mail has affected the world? How can we possible quantify how mobile mapping on your smartphone has benefitted society? It is not impossible, but it is taken for granted that this technology has, without a shadow of a doubt, saved billions of person hours. We do not spend hours trying to quantify the benefits of a technology; we rather take them for granted and expect the benefits to follow.

There is no tool that is more powerful than a well-crafted presentation that explains the value of a technology. GIS often fails to live up to expectations and, if misused, GIS can become a little more than a glorified mapping system. An analysis of the value, costs, and the perceived expected return is the only way to avoid these pitfalls. The investment in GIS within local government is strongly influenced by the level of commitment that is shown by elected officials. These officials understand the business case for technology, and this case is built on a thorough explanation and documentation of the VP of GIS technology. In my experience, an explanation of the value of the technology using real-world examples is, more often than not, followed by a commitment on behalf of decision makers to fund the initiative. The only way a GIS coordinator or manager can sustain this commitment is to continually explain to elected officials the value of this multipurpose technology.

8.8 The Value of a Life

Consider another important ramification of assigning dollar values to the benefits that the GIS offers. During a GIS strategic planning workshop, I demonstrated how the GIS potentially played a key role in saving a life. I showed how a 911 dispatch center used accurate, reliable, and real-time GIS data to pinpoint the location of an emergency situation. During the workshop, we discussed the value of GIS, and I asked the attendees to place a value on a human life. After all, we had documented evidence that GIS technology does save lives. Throughout all of the workshops where I have taught, not a single person has been forthright enough to assign a monetary value on a human life.

My response to this reticence is pragmatic: why can't we assign a human life a dollar value? After all, insurance companies do. For the sake of this hypothetical conversation, I am assigning a value of $1 million dollars across the board for a human life, regardless of his or her value to the society or any other metric. To do that, we need to develop a clear and understandable approach that starts not with a list of tangible and intangible benefits of GIS, but rather a discussion about the problems that are associated with presenting the value of GIS technology.

8.9 Life is Definable, Changeable, and Improvable

Let us start by examining what makes local government professionals and elected officials tick. There is a reason why our society is filled with self-help books. My frequent trips to the airport bookstore reveals a slew of books with titles like *Ten Ways to Manage People, Four Ways to Improve your Self Confidence, Nine Proven Techniques to Succeed,* and so on. On my mind, this is anecdotal evidence that we, as people, refuse to see ourselves in a bad light. We are filled with optimism. We have an unbelievable sense that life is definable, changeable, and improvable. We are not indifferent to our fate.

We may not care all the time but are made stronger by our desire to improve through the use of technology and the embrace of progress. I believe this to be uniquely American and a testimony to the greatness of this country.

Local government professionals all have a desire to improve operations. It may not work out all of the time, but the desire is unmistakable. To perform an effective assessment of the benefits that the GIS offers, we must review an organization's goals and action plans for improvement. What good would it be to prove that the GIS can effectively solve a problem that does not exist within a community?

A few years ago, I listened to a one-hour presentation on crime and crime analysis. The presenter carefully showed how the GIS could *save the day* with predictive modeling, instant real-time heat maps, and the ultimate changing of patrol beats to reduce crime. The problem was that this Alaska fishing port only had five crimes the previous year, all located at the same bar. Reducing crime was not a strategic initiative.

I may be playing the obvious game here, but it is vital that we understand an organization's needs before we demonstrate the utility of geospatial technology. GIS may be beneficial to virtually every area of local government, but you still need to focus on a specific organization's goals and objectives. Hot-button items are constantly changing. Believe me when I say that most GIS salespeople miss the mark by not listening to, or not taking the time to understand, the real needs of a municipality. If you do not know the needs or vision of an organization, how can you attempt to quantify the benefits of GIS? So, for heaven's sake, please take the time to understand an organization and review their *hot-button* items before you introduce the GIS.

8.10 A 70:30 Rule

My 70:30 rule is this: GIS can directly influence 70% of all the goals and objectives of an organization. If you are going to assess the VP or try to quantify the RoI of GIS technology, I recommend that you take a good, hard look at your organization's vision for the future.

Often, we have an unstructured and heavy-handed approach to the GIS. Sometimes, we force fit GIS. My guess is that most of you reading this chapter have not read and reread the vision, goals, and objectives of your organization. That is perfectly natural, but my recommendation is to read your organization's vision for the future and make sure that the GIS plays a supporting role. Remember that the GIS will not save the world itself; it is only a tool to help you save the world.

Let us look at the five major initiatives that are created for multiple organizations and begin the process of identifying the possibilities of GIS. Most organizations identify the distinctive characteristics that are important to that community and will direct their future growth. This can include, but are not limited to, the following five:

1. Quality of life
2. Economic diversity and innovation
3. Effective and efficient government
4. Infrastructure and facilities
5. Equity

My guess is that you read these initiatives and said to yourself, "Sure, I believe that GIS could support each of those initiatives." Before you look at the goals and objectives that follow, *stop*.

This is a test. Read each section and mentally describe how the GIS would support that goal and objective. For this test, I am using the 70:30 rule. That is to say that the GIS can be used to support at least 70% of every one of the goals that are listed as follows.

8.10.1 Quality-of-Life Goals with Supporting Objectives

Goal: Provide safe and reliable utility services
- Implement industry standard practices for the operation and maintenance of utility services, continually improving the overall system.
- Adhere to a capital replacement program to rebuild, replace, and upgrade utility infrastructure.

Goal: Maintain public safety
- Improve traffic safety and address community traffic concerns through education and law enforcement visibility.
- Engage in responsive and proactive code enforcement.
- Provide information about crimes and crime prevention.
- Utilize environmental design techniques to deter and prevent crime.
- Apply data drive analysis to develop strategies for more effective law enforcement and fire safety.
- Utilize data analysis to define public safety standards and evaluate law enforcement and fire service delivery models to enhance safety.

- Maintain and enhance disaster preparedness programs and ensure that the organization's emergency operations plan is current.

Goal: Develop and implement a strategy to increase the availability of housing choices

- Identify and promote sites for new housing development.
- Encourage diversity in housing products.
- Support in-fill housing projects and mixed-use development.
- Promote strong neighborhoods.

Goal: Establish planning and land-use practices that enhance the quality of life by promoting an orderly and balanced pattern of development and open spaces

- Identify underdeveloped and underutilized properties to encourage investment and new development.
- Encourage dense business nodes to reduce automobile traffic, promote walkability, and encourage alternative modes of transportation.

Goal: Provide quality open space, parks, and recreation

- Maintain and expand parks and trails.
- Maintain community parks and recreation assets.
- Seek opportunities to increase open space.
- Evaluate the feasibility of creating recreational/cultural activities utilizing existing open forest land.

Goal: Promote environmental sustainability

- Increase public awareness of and participation in conservation programs through education.
- Manage incentives and rebates for energy and water conservation.
- Maintain the commitment to environmental stewardship by improving energy efficiency, water and air quality, parks, open space, and urban wildland interface.

Goal: Work toward health in all policies by promoting healthy living and lifestyles

- Explore innovative and best management practices for promoting public health.
- Incorporate healthy living attributes into city-sponsored events.

Goal: Celebrate community diversity

- Develop and promote leisure activities for people of all ages, demographics, and cultures.
- Continue enhanced community building and engagement through technology and community events.

8.10.2 Economic Diversity and Innovation

Goal: Foster existing and create new infrastructures to support development

- Extend utility services.
- Evaluate adequacy of infrastructure to support future service needs.
- Cultivate economic diversity.
- Balance residential and commercial growth with quality of life.
- Allocate funds to support business recruitment efforts.
- Work to complete the economic asset inventory/study.
- Develop a business recruitment strategy based upon inventory results.
- Actively recruit businesses within desirable industries: (a) educational, (b) medical, and (c) technical.
- Encourage agriculture and food industries.

Goal: Support business retention and expansion efforts

- Support business retention and expansion.
- Work with the Chamber of Commerce to educate business operators on community planning and building requirements, including signage, business license, and other requirements.
- Expedite application approval and permit processing for business projects.
- Review commercial customer charges to identify conservation and best utility rates.

8.10.3 Effective and Efficient Government

Goal: Streamline the fundamentals of city operations

- Evaluate and update processes to ensure the application of best practices.
- Continue multidivision/department teamwork and shared resources/services without creating cross-subsidies.
- Create policies that provide clear guidance and assurance.
- Allocate and adjust internal resources to ensure success.
- Update business license application forms and integrate with processing software.
- Complete the enterprise resource planning replacement and expansion.

Goal: Utilize technology to enhance city operations

- Increase the use of technology.
- Standardize GIS to industry standard and build functionality that improves organization services.
- Plan for and fund the replacement of outdated software systems that provide efficiency and streamline government.
- Improve the citywide Wi-Fi system to improve mobile communication for field employees.

Goal: Increase public engagement, outreach, and communication

- Provide relevant and understandable documents and reports.
- Make council and commission meetings more interactive with real-time digital data.
- Hold an annual open house to enable citizens to meet with city council and staff and see technology at work.
- Offer a lecture series to educate citizens about city operations, community issues, etc.
- Improve the use of social media to provide information.
- Complete Website overhaul with public outreach focus on public engagement and awareness.

8.10.4 Infrastructure and Facilities

Goal: Address the deferred maintenance of organization facilities

- Evaluate the existing condition of all facilities and develop an objective rating system.
- Develop a maintenance program for organization assets based upon industry standards and manufacturer's recommendations.
- Use condition-based and value assessments to prioritize the repair, replacement, and construction of facilities.
- Implement a cost recovery program for organization-owned facilities that are used by others.

Goal: Develop a long-term capital replacement plan for public facilities.

- Inventory and provide valuation of public assets through GIS.
- Develop 2- and 10-year replacement plans for major capital projects.
- Use condition-based and solid economic analysis to prioritize repairs, replacement, and construction.
- Promote undergrounding of utilities through coordination with planned street reconstruction projects.

- Prioritize infrastructure needs based on availability of funds, life expectancy, and safety considerations.
- Effectively maintain ongoing infrastructure to new baseline levels.
- Efficiently and safely upgrade vehicles, utility infrastructure, and organization facilities with limited resources.

Goal: Install improvements to ensure the security of city facilities

- Install security fences, lights, cameras, and other equipment as necessary to protect facilities.
- Secure the city's water and wastewater systems.

8.10.5 Equity

Goal: Provide residents with the opportunity to improve their quality of life and well-being

- Develop, enhance, and monitor the organization's outstanding parks and recreation services.

By the time that you get to this section, you may have forgotten that the last 5 minutes was a test. How did you do? Did you achieve success in reaching 70%? Remember you were attempting to mentally describe how GIS would support each of the 5 goals of this organization.

8.11 An RoI and VP Solution

In early 2000, I spearheaded the development of what can only be classed as a straightforward, custom system for measuring the value of a GIS investment. Our system is more of an *approach*, rather than a system that is tailored to address the value of GIS implementation. This approach addresses successful outcomes but does not specifically address operational metrics such as response time or customer satisfaction. It measures 16 RoI categories or outcomes for local government. We deliberately used a simple system so that we could precisely define the process, measure the process, analyze the effectiveness, and then implement the recommended improvements. Some simplify performance measures into broad groups, including effectiveness, efficiency, quality, timeliness, productivity, safety, savings, and reduction.

The idea is to offer you something that you can use without extensive training or education. Here is the list of the tangible benefits of GIS, which I promised you. I supplement this list with a sublist of the risks that are associated with not implementing GIS. Let us call this list the 19 foundation key performance indicators (KPIs). You can build an RoI or a VP around this and explain what the risks of not doing anything are. Figure 8.1 sixteen RoI categories and Figure 8.2 RoI evaluation table graphically depicts this RoI assessment for local government.

8.11.1 KPI #1: Saving Money and Avoiding Costs

There is little doubt that GIS results in cost savings and cost avoidance. Immediate savings can be seen through better decisions and increased productivity. Cost avoidance becomes apparent as GIS helps organizations reduce and eliminate duplication of effort.

The risks associated with this key factor include the costs that are associated with the following:

- Risk of continuing to use outdated technology and techniques
- Risk of not using geospatial technology to streamline and improve operations

8.11.2 KPI #2: Saving Time

Having the information when you need and want it saves time, staff resources, and money. Information can be made available to the public through a Website or kiosk in convenient locations, reducing the demands on staff.

- Risk of continuing to duplicate work efforts and use outdated technology
- Risk of not streamlining and automating manual processes

8.11.3 KPI #3: Increased Productivity and Organizational Performance

Access to accurate, current information instantly saves local government staff from having to waste time searching for lost data or trying to correct inaccurate data. Accurate digital and electronic GIS mapping can be easily accessed by and shared among all departments.

- Risk of problems with the organizational culture
- Risk of not improving organizational performance, organizational sustainability, and accountability
- Risk of not having organization-wide resource planning

8.11.4 KPI #4: Improving Efficiency

GIS helps organizations reduce and eliminate redundant steps in workflow processes. GIS programs help reduce workloads and facilitate new procedures, resulting in increased productivity and efficiency.

- Risk of inefficiencies and redundancies
- Risk of not avoiding inefficient business processes
- Risk of not increasing productivity through new technology

8.11.5 KPI #5: Improving Data Accuracy and Reliability

The GIS creates maps from data, or paper maps can be digitized and translated into the GIS. Maps can be created on any location, at any scale, and show selected information to highlight specific characteristics. Precise GIS data enable users to generate accurate reports and produce quality maps instantly.

- Risk of inaccurate information, maps, and statistical reports

8.11.6 KPI #6: Making Better and More Informed Decisions

The GIS is a critical tool to query, analyze, and map data in decision support. The GIS can, for example, be used to choose a location for a development that has minimal environmental impact, is located in a low-risk area, and is close to a population center.

- Risk associated with damaging the reputation of the organization or undermining confidence in the organization
- Risk of assumption-based decision making

8.11.7 KPI #7: Saving Lives and Mitigating Risks

In an emergency, when every second counts, the GIS can lead rescuers quickly and accurately to the scene. The time saved in locating a citizen can be the difference between life and death.

- Risk that employees are negatively impacted or physically harmed

8.11.8 KPI #8: Automating Workflow Procedures

GIS helps automate tasks that expedite workflow and enhance your ability to react efficiently during a crisis. The GIS can automate routine analysis, map production, data creation and maintenance, reporting, and statistical analysis.

- Risk of information hoarding and missing information
- Risk the inability to locate critical or timely information

8.11.9 KPI #9: Improving Information Processing

Enterprise-wide GIS streamlines the flow of information throughout the organization, leading to better accuracy, better access, and increased efficiency in every aspect of the organization.

- Risk of continued data and process duplication

- Risk that there will be poorly maintained, misplaced, and stale information
- Risk of not having easy geographic exchange

8.11.10 KPI #10: Complying with State and Federal Mandates

Digital inventories of water, sewer, and storm water infrastructure are becoming increasingly important in local governments. A complete GIS program includes asset management, inventory control, and depreciation based on accurate and timely data including age, size, and construction materials; this allows managers to predict and schedule repairs and replacement.

- Risk related to the consequences of noncompliance with laws, regulations, and policies

8.11.11 KPI #11: Protecting the Community

The GIS helps public safety officials to develop emergency plans and respond to disasters more effectively than ever before. It provides tools to monitor conditions, recognize threats, predict consequences, and respond effectively and efficiently to man-made or natural disasters. It can also help officials deliver information to citizens during an emergency, through emergency notification systems and the Internet.

- Risk that citizens negatively impacted or physically harmed
- Risk the inability to prevent and respond to an emergency situation

8.11.12 KPI #12: Improving Communication, Coordination, and Collaboration

Good communication is the key to running an effective organization. The GIS helps staff and elected officials convey complex information in easy-to-understand formats.

- Risk that there will continue to be variations in priorities
- Risk that there will be inefficient decision making
- Risk of poor training and education
- Risk of insensitivity to user needs
- Risk on everyone doing their own thing and going their own way
- Risk of not working as a team
- Risk a lack of strategic decision making
- Risk a lack of stakeholder consent building
- Risk clear lines of responsibilities and accountability

8.11.13 KPI #13: Provide Data to Regulators, Developers, and Other Interested Parties

Making digital data available to regulators, developers, and other interested parties via the Web or by data storage device streamlines and ultimately improves services to the community. Making GIS data available to interested parties will save the organization money.

- Risk of duplicating effort, time, and resources
- Risk of hoarding data and information
- Risk on wasting money on data duplication

8.11.14 KPI #14: Respond More Quickly to Citizen Requests

GIS data support intelligent and timely response to citizen requests. Timely information about crime, permits, emergency situations, and general land information, including zoning and land use, improved efficiency.

- Risk of not meeting customer expectations
- Risk timely response to citizen requests
- Risk of ill-informed citizens

8.11.15 KPI #15: Improve Citizen Access to Government

GIS-centric Internet solutions and crowdsourcing applications make citizen access to digital data streamlined and rapid. Immediate access to critical data is made available through GIS.

- Risk that citizens will be negatively impacted
- Risk to open and transparent democratic government
- Risk of misinforming the public
- Risk of not seeing GIS information as a public resource

8.11.16 KPI #16: Effective Management of Assets and Resources

The effective management of a local government's infrastructure assets starts with GIS. Tracking, analyzing, managing, and conserving assets are key components of asset management.

- Risks related to decisions about assets such as liabilities, asset management, capital and operational funding, and economic development
- Risk that there will be internal competition over funding projects and resources
- Risk of project and process management

The following three additional KPI's were added during some recent strategic planning projects.

8.11.17 KPI #17: Good Environmental Stewardship and Well-Being

GIS is being used in many ways to promote community well-being, healthy populations, environmental protection, community vitality, leisure, and cultural education.

- Risk that the physical environment will be damaged
- Risk of not promoting the well-being of citizens
- Risk of not promoting healthy populations
- Risk of not managing parks, leisure, and cultural education

8.11.18 KPI #18: Data Relationship—New Ways of Thinking

GIS is offering new ways to think about data. It turns data into meaningful information. New relationships are being developed between data, information, and decision support. Predicting where and when crime will occur, and the relationship or link between arson crimes with other crimes, is a new way of data interpretation. The relationship of the age and condition of infrastructure and its life cycle are examples of improved spatial thinking, as well as new data relationships.

- Risk of not exploring how we can make our organization better by analyzing data and turning them into meaningful information
- Risk of not using technology to improve decision support and forecasting

8.11.19 KPI #19: Promote Economic Vitality

Local government organizations are taking advantage of how GIS can give them a competitive advantage. Attracting new business and retaining existing business promotes economic vitality. Innovative use of GIS technology that provides easy-to-understand information is becoming invaluable.

- Risk of not promoting the economic opportunities of the organization
- Risk of not using digital GIS data effectively
- Risk of not using new geospatial tools to promote local business opportunities
- Risk of not promoting the organization as a high-tech community

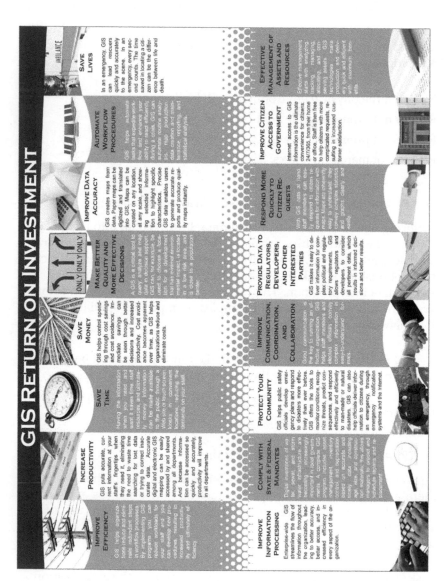

FIGURE 8.1
Sixteen RoI categories.

RETURN ON INVESTMENT OPPORTUNITIES	DEPARTMENTS																									
	Airport	Attorney's Office	Board of Elections	Building Permits & Inspections	Central Services	Computer Information Services	Cooperative Extension	Economic Development	Emergency Management	Finance	Fire & Emergency Services	Housing & Community Development	Human Resources	Leisure Services	Manager's Office	Planning	Police	Probation Services	Public Information Office	Public Utilities	Sheriff's Office	Solid Waste	Tax Assessor's Office	Tax Commissioner's Office	Transit Department	Transportation & Public Works
Automate Workflow Procedures				●		●				●						●				●						
Comply with State & Federal Mandates											●						●			●						●
Effective Management of Assets & Resources	●			●	●			●	●		●	●		●		●	●	●		●		●	●	●	●	●
Improve Citizen Access to Government			●				●	●	●	●		●		●		●	●		●					●	●	●
Improve Communication, Coordination & Collaboration	●	●	●	●	●	●	●	●	●	●	●	●	●	●	●	●	●	●	●	●	●	●	●	●	●	●
Improve Data Accuracy				●	●						●	●		●		●	●			●			●		●	●
Improve Efficiency	●		●	●	●			●	●	●	●	●		●	●	●	●	●	●	●	●	●	●	●	●	●
Improve Information Processing						●										●										
Increase Productivity	●		●	●	●			●	●	●	●	●		●	●	●	●	●	●	●	●	●	●	●	●	●
Make Better Quality & More Effective Decisions	●	●	●	●	●			●	●	●	●	●	●	●	●	●	●	●	●	●	●	●	●	●	●	●
Protect Your Community									●		●						●	●								
Provide Data to Regulators, Developers & Other Interested Parties				●				●				●		●						●			●	●	●	●
Respond More Quickly to Citizen Requests			●		●	●	●	●	●		●	●		●	●	●	●		●			●	●	●	●	●
Save Lives									●		●						●									
Save Money		●	●	●	●	●		●	●	●	●	●	●	●	●	●	●		●	●		●	●	●	●	●
Save Time	●	●	●	●	●	●	●	●	●	●	●	●	●	●	●	●	●	●	●	●	●	●	●	●	●	●

FIGURE 8.2
RoI evaluation table.

Figure 8.1 shows the RoI categories that illustrate a simplified way of looking at the RoI of an organization. It is a simple and effective map of all opportunities.

Figure 8.2 illustrates how to quickly and easily identify the value of GIS for each department within local government. It is a simple but effective tool that will allow you to identify a value and then begin the process of detailing the RoI.

9

How to Sell GIS to Local Government

It has to be within you, as desire for drink is within a drunkard, or love is within the lover.

Henri Matisse

9.1 Introduction

The answer to the question "How do you sell the GIS to local government?" may be simple, but it took me 20 years to figure out. What I am about to say may sound corny to some of you, but here it goes: I believe that GIS is not a gadget but something that will ultimately transform the way that people think about the world. That is why I am so comfortable selling GIS to local government organizations.

People do not buy what you do; they buy what you believe, that is, they buy the idea of thinking differently. People buy from people who enhance their belief that change is possible. If you truly believe in something, others—clients, employees, friends—will too.

Local government professionals love to think out of the box. They love to think differently. Being a good steward of taxpayer's money includes mastering the art of change and, along with this, new business cultures. Just give them an opportunity. Success in selling comes when you have a genuine belief in GIS, and you truly know why. This may be finding your bliss. It is not a job; it is a passion. There is no incentive that is more powerful than an idea or a belief. Many great leaders somehow inspire others through a belief system.

I do believe in the *why–how–what strategy* to life and sales. I also believe that the why question is the most fundamental component of this strategy and warrants consideration above all else. Great leaders understand deep within themselves why they do what they do. Simon Sinek, author and best known for popularizing the concept of *the golden circle*, calls this the golden circle of why, what, and how, and it is a compelling argument for all of us to act upon what we believe in. People buy what you do because they can see why you do it.

9.2 GIS Is Immensely Sound and Rich with Benefits

A wise person once said, "Communication is a key component to most of life's successes." I have never regarded myself as a salesman, but I do feel that my straightforward storytelling ability may just qualify me for that category. I am not much for *cold calling*, but give me an audience of decision makers and the task of justifying GIS technologies, and I am off to the races. I do not have a very good memory, nor am I particularly articulate, but that being said, I do not believe that quantifying and qualifying the benefits of GIS is a terribly difficult task. On the contrary, *GIS is immensely sound and rich with benefits. It is laden with extraordinary examples of how it supports our world. It has made and continues to make a dramatic and positive difference in the way that local governments operate in the twenty-first century.*

The objective of this chapter is to explain how to sell GIS within the local government context. I believe that a GIS coordinator of any worth should, within a moment's notice, be able to sell GIS to every department within the organization. Can most GIS coordinators do this? Probably not. So, if you have ever said to yourself, "I don't know how to sell GIS to my organization or my elected officials," or if you are a consultant or academician who is interested in one man's approach to selling to local government, then this chapter may be the one for you.

At this early point in this chapter, I want to make clear that I am not really talking about *sales*. I am talking about *selling*, where you intelligently articulate a case for GIS. There seems to be a distinct lack of literature that is related to selling GIS, especially selling to local government, and with this chapter, I aim to change that.

In the next few sections, we will discuss what GIS is, and why local governments invest so heavily in it. We will consider both the *obvious* and the *hidden* forces that shape how *decision makers* think, feel, and behave and how these apply to the selling of GIS technology. The next few sections are devoted to these *forces*, in addition to the provocative language of selling and the importance of astute word choice. This chapter will detail the role of the GIS coordinator as a mechanism for selling GIS and document the critically important art of listening and processing stakeholder needs.

The following sections also describe the benefits of aligning your sales pitch with the organization's overall vision, goals, and objectives, hence the term *alignment*. I take a leaf out of the business world and discuss how *branding* a GIS initiative can improve its perceived ability to meet departmental business needs.

Finally, I offer a clear and concise discussion of the issues that are related to local governments recovering the cost or even generating revenue from the creation and dissemination of GIS data for both public and private consumption.

So, this leaves us with six distinct topics that require serious discussion:

1. What exactly is GIS?
2. Why do local government organizations invest in GIS?
3. The forces that shape how we think, feel, and behave.
4. The provocative language of selling and framing your argument.
5. A loophole in our sales strategy.
6. Language and sales.
7. Selling local government data.

This preceding list is either absurdly ironic or incredibly poignant because my real interest in sales is the hidden forces that shape the thoughts, feelings, and behavior of the people who buy into the concept of GIS. The accidental irony is that the number seven or the *power of seven* is particularly important in marketing and selling. It is about memory. The power of seven (plus or minus two) is argued to be the number of objects that an average human can hold in working memory (Miller's law—cognitive psychologist George A. Miller of Princeton University's Department of Psychology). This is why our telephone numbers are seven digits. This is a fact that we can apply practically to our sales strategies.

9.2.1 Selling Topic One: What Exactly Is GIS?

Before we embark on this psychological guide to selling GIS technology to and within local government, it is important to explain what we are selling. GIS means many things to many people. Even though I have tried desperately not to do what most authors have done over the past 25 years, namely, develop one simple paragraph that explains GIS, I am about to do it. So, with that being said,

> GIS is a computer system that is designed to allow "people" to capture, store, assemble, analyze, manipulate, and display geographic and location information.

What a horrible and old-school definition of a GIS. What if we added this statement?

> GIS is used by people to investigate, plan, manage, evaluate, calculate, predict, and assess our world.

It sounds better and more interesting but still needs to substitute the traditional statements with examples of what the technology currently does.

> GIS is used by virtually every person in the developed world to navigate, locate, engage, change, and model our multidimensional world.

What if we describe it purely in terms of a local government GIS? After all, this is a book about a formula for local government GIS success. Here goes my definition of local government GIS:

> GIS is a proven organization-wide, enterprise, and enduring technology that continues to change how local government operates. It defines people's relationship with geographic components and analytics. It makes a government stronger with its immense potential to improve efficiency, innovate existing outdated practices, manage assets, increase productivity, offer real-time decision support, and navigate and predict future events. GIS is immensely sound and rich with benefits. It is laden with extraordinary and wonderful examples of how it supports our world. It has made and continues to make a dramatic and positive difference in the way that local government thinks and operates.

GIS is one of the only technologies to have the power to enhance the quality of life, promote a healthy and clean environment, guarantee good, sound infrastructure, and endorse social equity. Notice that I include people or persons in all my explanations. Your job is to explain why local government would be poorer for not investing in GIS. Figure 9.1 documents the key components of an enterprise GIS.

FIGURE 9.1
The components of a local government GIS.

- GIS governance
- GIS digital data and databases
- Procedures, workflow, and integration
- GIS software
- GIS training, education, and knowledge transfer
- Infrastructure

9.2.2 Selling Topic Two: Why Do Local Government Organizations Invest in GIS?

The next question we need to answer is, why do local government organizations invest in GIS? If we know why they purchase GIS solutions, we are much closer to developing a sales strategy.

We should bear in mind that a local government organization's investment in GIS is usually substantial and often requires thought and planning. So, what are the major reasons behind this substantial purchase?

Let us agree that the factors influencing the purchase of a desktop GIS by the public works department are far different in nature from the factors influencing a million-dollar investment in an enterprise solution. The following list in Figure 9.2 is based on my local government experience and outlines all of the possible reasons why organizations invest in an enterprise GIS.

The following seven key factors (remember the power of seven), listed in Figure 9.3 in order of priority, are the critical reasons that many organizations invest in or adopt the GIS.

REASONS WHY LOCAL GOVERNMENT INVEST IN GIS

- ☑ Organizational staff – Key People
- ☑ The development of a GIS strategic implementation plan
- ☑ A good sales pitch from a vendor or consultant
- ☑ True and legitimate organizational needs
- ☑ True and legitimate departmental needs
- ☑ An enterprise need
- ☑ Distinct operational needs (Parcel Management, Work Order Solution or Crime Analysis)
- ☑ Culture and tradition of technology
- ☑ Politics
- ☑ Cost benefit analysis
- ☑ Available funds
- ☑ Timing
- ☑ The Idea of GIS
- ☑ Existing Success
- ☑ The GIS Team

FIGURE 9.2
The many reasons why local governments invest in GIS.

FIGURE 9.3
The top seven key factors why local governments invest in GIS.

There is not any empirical evidence to support my seven-factor theory, but as a longtime member of the industry, I have seen these factors crop up countless times. It is easy to develop a list of the reasons why GIS is acquired. It is harder to understand our audience and the forces that shape what and how they think; however, this is fundamental to our sales strategy. The power of seven is a nice and easy introduction to the one facet of the human condition—namely, that we cannot remember stuff. This is a subject that is completely ignored in the world of sales and GIS.

Section 9.2.3 will help you understand the forces that shape people's decisions, whether they be the obvious forces or the hidden, ignored ones that ultimately play a significant role in determining the success of your sales pitch.

Though there are certain shared characteristics, sales and marketing conducted in local government are different from the way that they are conducted in the standard business world.

So let us discuss the components of selling GIS within local government. The following sections will show you how to convince all stakeholders, elected officials, and decision makers that GIS should be their technology of choice.

Before going forward, I would like to note that I just secured a $410,000 GIS adoption project with a native Alaskan corporation. Tell me, how did a 53-year-old Yorkshire man sell GIS to a native tribal corporation?

The answer is understanding your audience and believing in the recommended solution.

9.2.3 Selling Topic Three: Forces That Shape How We Think, Feel, and Behave

Never underestimate the forces that shape how we think, feel, and behave. First, there are the *obvious forces* that shape the way that elected officials and decision makers think about GIS. For example, if you are presenting empirical evidence that proves beyond a shadow of a doubt that GIS will benefit the organization, or illustrates how other organizations' best business practice (BBP) evidence shows the benefits of GIS technology, or reliably benchmarking your organization against other similar organizations, then you are building a case for increased GIS utilization by addressing the obvious forces that will influence the *buyers* who are, in this case, elected officials.

Sales is all about *shaping perception*. Remember how I told you that people buy from people? As they say, *perception is the reality*. People will believe what you believe, if given enough incentive.

However, the real question you have to ask yourself is, are you using all of these obvious forces to your benefit? We will detail every obvious force that shapes perceptions in detail later in this chapter.

Of course, if there are obvious forces, this means that there are also *hidden forces* that affect how and whether elected officials or senior staff decide to support a GIS initiative. These hidden forces are not to be taken lightly or, as is too often the case, completely ignored. If we want to talk seriously about selling the GIS in local government, these forces require address.

A chief hidden force is *alignment* and pertains to questions about how GIS technology aligns with the organization's existing mission, vision, goals, and objectives. These objectives often entail improving the quality of life, maintaining a healthy and clean environment, innovation, maintaining sound infrastructure, and increasing social equity.

Hidden forces can also include the political agendas of a council or just the personal preferences of the *buyers*. The *maven* factor can also be a hidden contribution to sales. The maven is an individual who causes a butterfly effect of word-of-mouth support for GIS technology. Some other hidden forces include marketing tools and *tricks of the selling trade*.

It is not always a level playing field when it comes to sales. One thing we must agree on is the power of the obvious forces and the hidden or ignored forces, especially those outside of our control. We must acknowledge that it is not always a level playing field when it comes to sales. There are things happening beyond our control that must be considered. Understanding the things that motivate and affect people will set your GIS sales expertise in motion.

For example, how much control do elected officials have over their decisions? An easily recognized force is the lack of profit motive in government. For most local government officials, it is about providing the citizenry with the best possible services for the least possible cost. It is about the effective management of taxpayer's dollars, a democratic process, and good stewardship of funds. These foci tend to make governments risk aversive.

Do not, however, underestimate the ego factor. Senior officials in local government want to improve and change their communities and may sometimes see GIS adoption as the legacy that they will leave.

Remember, as we go forward, we are not just talking about selling GIS cold to an organization. Most of the time, we are in a situation where a GIS strategic plan has been completed and the three-year funding needs to be ratified. Both the obvious and the hidden forces play a significant role in determining your success in selling the GIS to the organization. In the world of selling, we must embrace both the obvious and the hidden forces that affect the people who are making the decisions. Figure 9.4 illustrates these forces and how they affect GIS implementation.

Let us take a good look at all of these forces as listed in Figure 9.4.

Forced that Shape and Influence Decisions	
OBVIOUS FORCES	**HIDDEN FORCES**
• Empirical Evidence	• Political Agenda
• Quantifiable Results	• Personnel Preferences
• Anecdotal Stories	• Exposure to Images
• Demonstrations	• Language
• Consensus	• Alignment
• Case Studies	• Environment
• Best Business Practices	• Maven/Word of Mouth
• References	• Body Language
• High Tech Solutions	• Open Government
• Benchmarking	• Code of the Selling Trade
• Blog/Newsletter	• Ego
• Awards	• Intangibles

FIGURE 9.4
The obvious and hidden forces that affect the sale of GIS.

9.2.3.1 Obvious Forces

- *Empirical evidence*

  This is information that is gathered by observation or experimentation. This source of knowledge can come from firsthand experience in the field or case studies that are documented in GIS journals and publications.

- *Quantifiable results*

  This is a logical, explicit expression of the quantity of something. It refers to items that are capable of being measured or counted. They say, if it cannot be measured, it does not exist. Keep this in mind when it comes to selling the GIS.

- *Anecdotal stories*

  An anecdote is a short, often-amusing account of events that are related to GIS activity. Anecdotal evidence is not usually published but serves as a way to illustrate the lighter side of GIS implementation. Think of anecdotes as GIS parables. Though this type of evidence can be unreliable, anecdotes are particularly useful in the hands of a master salesperson.

- *Demonstrations*

  This is the act of showing others how something is used or done. A GIS software demonstration proves to users that the software functions in a specific way. Live demonstrations are particularly important for the skeptics.

- *Consensus*

  Consensus is an agreement among decision makers, GIS users, and elected officials. Consensus is important, as it shows agreement and harmony in moving forward with GIS implementation.

- *Case studies*

  A case study is often used to present the BBPs for the GIS initiative. It can also illustrate the benefits and success stories of GIS implementation. A case study is a record of research that expresses the results that are gathered over a period of time.

- *BBPs*

  This is a method, technique, or process that allows for the development of standard operating procedures (SOPs) that can be used as a benchmark. In short, this is the best way for a business to accomplish something.

- *References*

  A reference describes another organization's successful approach to GIS planning, design, and implementation. During a presentation, it is always useful to reference successful organizations. Remember that if, during your sales pitch, you are going to reference other organizations, please try to make sure that they are of a similar size and complexity.

- *High-tech solutions*

  Innovative scientific methods that have a *wow factor* can often be an important part of a sales pitch or presentation. These show the ways that cutting-edge technologies introduce a new way of thinking or an automated way to accomplish a traditionally manual process.

- *Benchmarking*

  Benchmarking is a way to measure something by comparing and contrasting it with an accepted standard. This can often be the measurement of an organization's services, solutions, programs, strategies, and GIS implementation.

- *Blog/newsletter*

  Promoting success through social media has become commonplace and should be an important part of government operations. Blogs or newsletters are communication tools that promote GIS activity, success, and tips and techniques. Presenting improved communication tools is an important part of a sales pitch.

- *Awards*

  A mark of recognition given in honor of achievement is a goal that all local government organizations aspire to. Mentioning the potential for an organization to win awards incentivizes decision makers to implement GIS technology.

9.2.3.2 Hidden Forces

- *Political agenda*

  The political agenda is a set of issues or policies that are laid out and pursued by an individual or a group that is interested in furthering that idea. Political agendas are often hidden from the salesperson. It is useful to attempt to understand an elected official's agenda. It may have a lot to do with alignment.

- *Personal preferences*

  Decision makers are human and thus are influenced by an internal encyclopedia of factors. Personal preferences can be changed if a solid sales argument is demonstrated. Remember that decision makers may favor a specific software solution because of a recent conference or presentation.

- *Exposure to images*

  Rich graphics, supported by meaningful maps, will significantly impact decision makers. The new and innovative ways of mapping data are a solid approach to sales. When it comes to selling GIS, a picture is really worth a thousand words. However, you must have a wow factor.

- *Language*

  The sincere language that hits the mark and is supported by an intelligent vocabulary is an essential ingredient in sales. Positive sales words are important, but be sure to use layperson's terms instead of lingo and technical jargon that may confuse or turn off buyers.

- *Alignment*

  Alignment refers to the art of talking about the benefits of GIS in the context of exact terms that are laid out in an organization's vision, goals, and objectives. There is a great advantage in calling upon the moral framework of the decision makers when discussing alignment.

- *Environment*

  The environment is the atmosphere and ambience of the situation. Usually, the sales environment involves a meeting with council members, but it could be anything up to a dinner with elected officials. Your job is to create a welcoming environment by demonstrating how this GIS technology will make the organization more efficient. By all means, assert yourself, but never ever let the environment spiral out of control by becoming confrontational or dismissive.

- *Maven/word of mouth*

  A maven is an expert who knows a good deal about the GIS and spreads the word to colleagues. Never underestimate the word-of-mouth factor. Local government officials have a solid network. This should work to your advantage if you are driven to excel. Take it one client at a time. Encourage the maven factor.

- *Body language*

  The body expresses confidence, hesitation, and truth in subtle ways. The question you should always ask yourself is, what is my body language saying to decision makers? They say that you can predict the likelihood of a doctor being sued, just by watching his or her body language. Therefore, good posture is critically important, as are firm handshakes. Do not fold arms and look down and to the right. If you feel powerful, you will become powerful.

- *Open government*

  Open and transparent is always a buzzword in local government. Use it. *Open government, open data* fits this description and refers to sharing data with all interested parties. Publishing data and sharing information improve collaboration and communication between decision makers, staff, and citizens. There are guidelines for what constitutes open government data.

- *A code of the selling trade*

   Your personal code of the selling trade refers to your fundamental sales compass. What is this sales compass? It is a key list of things that encircle everything you do and can never forget. This list will include understanding people and why they do things on an emotional rather than logical level, understanding the characteristics of good leaders, and approaching other people with humility and respect. This is your list.

- *Ego*

   We all have egos. Elected officials and other decision makers within local government are no different. An ego is a person's sense of self-esteem or self-importance, and this factor often plays a role when selling GIS to local government. Feelings of self-importance often go hand in hand with a political agenda.

- *Intangibles*

   This is something that is difficult to measure. It is something that cannot be touched or grasped and has no physical presence. GIS brings many intangibles to the table, but so do people. For example, interpersonal goodwill is an intangible factor. Presenting the intangibles is useful and should be done responsibly.

Figure 9.5 illustrates the power of the obvious and hidden sales forces that influence the decision to implement enterprise the GIS in four organizations. The four organizations include the following:

1. City in Alaska
2. Town in California
3. City of Georgia
4. City of Alabama

In Figure 9.5, each column describes which *force* influenced the decision to invest in enterprise GIS. The forces are numbered 1 to 5 according to my estimation of which forces were the priority. Number 1 denotes the highest priority, and number 5 denotes the least. All I am trying to do here by referencing some good clients is to show you that you should always consider each obvious and hidden force.

9.2.3.3 A Values Game

During the writing of this chapter, my wife brought home a *values game*. This was a particularly beneficial coincidence as it illustrated how difficult it is

THE FACTORS THAT INFLUENCE THE DECISION TO INVEST IN GIS

Organizations »	City in Alaska	City in California	City in Georgia	City in Alabama
Obvious Forces				
Empirical Evidence		4	4	
Quantifiable Results				
Anecdotal Stories				
Demonstrations		5	5	
Consensus	2			
Case Studies				
Best Business Practices	3	1	2	1
References				
High Tech Solutions	1	2		2
Benchmarking			1	
Blog/Newsletter				
Awards				
Hidden Forces				
Political Agenda			3	
Personnel Preferences		3		3
Exposure to Images				
Language				4
Alignment	5			
Environment				5
Maven/Word of Mouth	4			
Body Language				
Open Government				
Code of the Selling Trade				
Ego				
Intangibles				

FIGURE 9.5
The factors that influences the decision to invest in GIS.

to predict or even accurately guess the values of individuals. We, as sales-people, would subscribe to Roy E. Disney's statement that "It's not hard to make decisions once you know what your values are." My point is that, even after talking to people, it is still very difficult to know their values. Play the following game to see what I mean.

The Values Game

The objective of this game is to identify your top five values, and by doing so, it will help guide your decision making. We are interested in one point and one point alone: the incredible difficulty of predicting people's values. There are three categories to consider:

1. Always valued
2. Often valued
3. Sometimes valued

Figure 9.6 documents the components of the game. There are 40 values as listed in Figure 9.6.

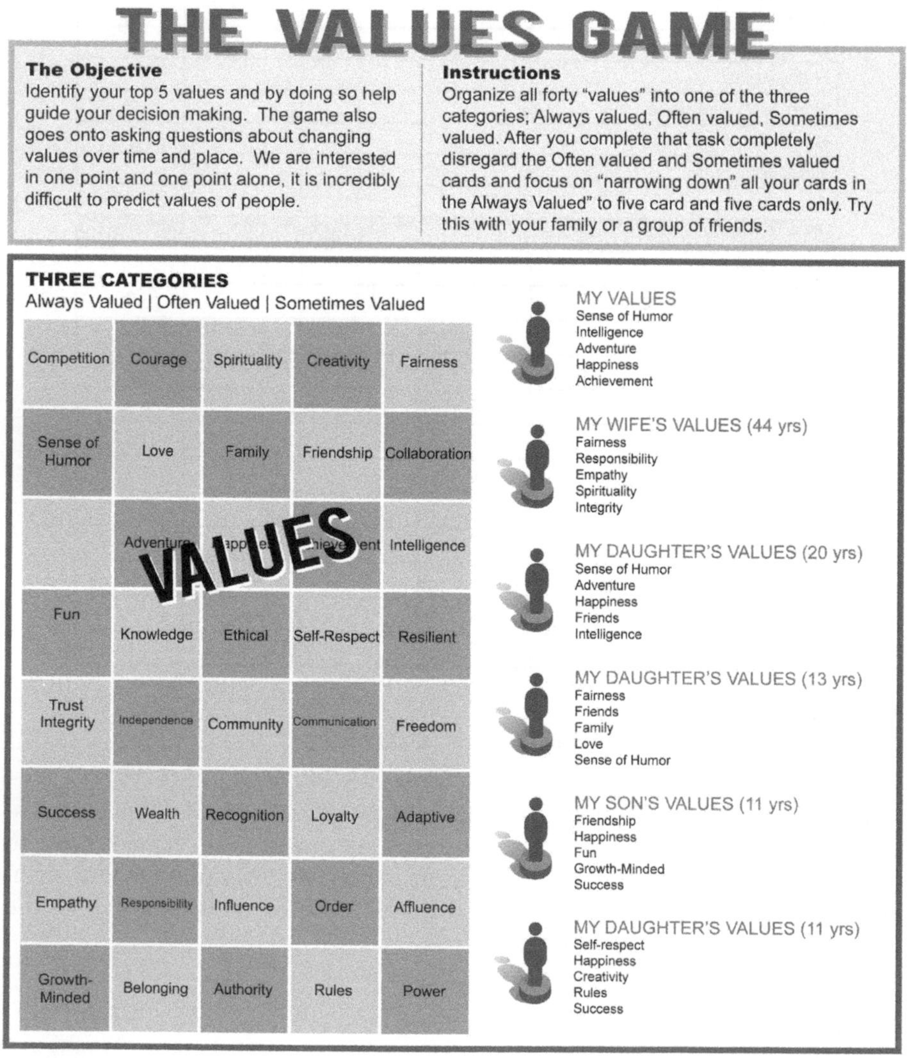

THE VALUES GAME

The Objective
Identify your top 5 values and by doing so help guide your decision making. The game also goes onto asking questions about changing values over time and place. We are interested in one point and one point alone, it is incredibly difficult to predict values of people.

Instructions
Organize all forty "values" into one of the three categories; Always valued, Often valued, Sometimes valued. After you complete that task completely disregard the Often valued and Sometimes valued cards and focus on "narrowing down" all your cards in the Always Valued" to five card and five cards only. Try this with your family or a group of friends.

THREE CATEGORIES
Always Valued | Often Valued | Sometimes Valued

Competition	Courage	Spirituality	Creativity	Fairness
Sense of Humor	Love	Family	Friendship	Collaboration
Adventure	Happiness	Achievement	Intelligence	
Fun	Knowledge	Ethical	Self-Respect	Resilient
Trust Integrity	Independence	Community	Communication	Freedom
Success	Wealth	Recognition	Loyalty	Adaptive
Empathy	Responsibility	Influence	Order	Affluence
Growth-Minded	Belonging	Authority	Rules	Power

VALUES

MY VALUES
Sense of Humor
Intelligence
Adventure
Happiness
Achievement

MY WIFE'S VALUES (44 yrs)
Fairness
Responsibility
Empathy
Spirituality
Integrity

MY DAUGHTER'S VALUES (20 yrs)
Sense of Humor
Adventure
Happiness
Friends
Intelligence

MY DAUGHTER'S VALUES (13 yrs)
Fairness
Friends
Family
Love
Sense of Humor

MY SON'S VALUES (11 yrs)
Friendship
Happiness
Fun
Growth-Minded
Success

MY DAUGHTER'S VALUES (11 yrs)
Self-respect
Happiness
Creativity
Rules
Success

FIGURE 9.6
Values game.

Organize all 40 *values* into one of the three categories: (1) always valued, (2) often valued, or (3) sometimes valued. After you complete that task, completely disregard the often-valued and sometimes-valued cards. This was obviously a red herring. Narrow down all your cards in the always-valued category to five cards and five cards only. Try this with your family or a group of friends, as I did. The results of my family are as follows, and they will give you a sense of what to expect:

My values

- Sense of humor
- Intelligence
- Adventure
- Happiness
- Achievement

My 44 years old wife's values

- Fairness
- Responsibility
- Empathy
- Spirituality
- Integrity

My 20 years old daughter's values

- Sense of humor
- Adventure
- Happiness
- Friends
- Intelligence

My 13 years old daughter's values

- Fairness
- Friends
- Family
- Love
- Sense of humor

My 11 years old son's values

- Friendship
- Happiness
- Fun
- Growth minded
- Success

My 11 years old daughter's values
- Self-respect
- Happiness
- Creativity
- Rules
- Success

What does this tell us?

The only thing we can conclude from these findings is that it is extremely difficult to determine people's values. Even your own family may surprise you. The only advice I can give you is to read and understand the values of the organization that you work for and use them to leverage GIS.

To know what people are thinking before they even think of it. If I could do it all again, I would consider studying psychology. After all, this discipline is the study of the mind and human behavior. Psychology seeks to understand individuals' attitudes, mind-sets, and feelings. Can you imagine having the ability to know what people are thinking before they even think of it? Wow! If you could actually have known what those 12 council members were thinking before you asked them for $1.2 million for the three-year GIS budget, or what the information technology director was thinking when you introduced the idea that GIS needs to be expanded throughout the organization and will require more time, resources, and budget?

I will openly admit that when it comes to being *psychic*, I have zero powers, and I am poorer for it. I was once told that one of my grandmothers had the *power*—the power being the ability to predict the future—but sadly, she never mentioned anything to me about gene sequencing, or in-car navigation, or that Blockbuster would completely disappear. Sadly, she took her powers to the grave.

Close your eyes and project yourself into the geospatial future. Visualize the future. Did it work? I am guessing that, like me, you came up with very little. So, it would seem that we are left to use a combination of common sense and the results of psychological experiments. Then, we have to explain these in layperson's terms.

Recently, I have become extremely interested in books (e-books) and television programs (Netflix) that reveal how our feelings, thoughts, and decisions are all influenced by forces beyond our control. I have watched every episode of the *Brain Games* and have read *Drunk Tank Pink and Other Unexpected Forces that Shape How We Think, Feel and Behave* (2013) by Adam Alter, *Stumbling on Happiness* (2006) by Daniel Gilbert, and *Predictably Irrational and Other Hidden Forces that Shape Our Decisions* (2008) by Dan Ariely.

One of my favorite quotes is from Alter is that "Humans are biologically predisposed to avoid sadness, and they respond to sad moods by seeking opportunities for mood repair. In contrast, happiness sends a signal that everything is fine, the environment doesn't pose an imminent threat, and there's no need to think deeply and carefully" (pp. 219–220).

This statement underscores that fact that successful buying and selling in itself is at the discretion of the human spirit. Let us take a deeper dive into the world of the forces that will shape how people perceive the GIS and ultimately purchase a solution.

9.2.4 Selling Topic Four: The Provocative Language of Selling and Framing the Argument

On National Public Radio's Hidden Brain Podcast (12/17/15) Shankar Vedantam asked the question, "Is arguing with passion the most effective way to persuade other people or opponents?" The answer, based on research from the University of Toronto, Canada, states that "Our unconscious moral framework shapes how we make arguments. Research indicates that if you want to persuade people, you should frame your points using your opponents' moral framework" (Podcast). This is exactly what I am saying when I tell you to make sure that your sales pitch is in alignment with your organization's own vision, goals, and objectives.

The reason this idea interests me is that when we talk about selling GIS to local government, we more often than not miss the *moral framework* factor that includes the following:

- Aligning the sales pitch with the vision, mission, and goals and objectives of the organization
- Framing our sales pitch in the moral framework of decision makers

The table in Figure 9.7 illustrates the moral framework of a GIS coordinator and an elected official. Please note how different they are from each other.

Understanding our Moral Framework	
GIS COORDINATOR MORAL FRAMEWORK	**ELECTED OFFICIAL MORAL FRAMEWORK**
• Buffer	• Equity - all citizens
• Intersect	• Economy - strong growth
• Overlay	• Fiscal Responsibility
• Calculate	• Improve health of citizens
• Statistics	• Safer society
• Analyze	• Stronger organization
• Predict	• Provide a cleaner City
• Interpret	• Provide for a well-run City
• Boolean Logic	• Provide citizen access to City data
• Polygon	• Allow the City to provide quality public service
• Point	
• Line	• Empower stakeholders
• Visualize	• Promote good steward of conservation
• Databases	• Engage and notify citizens
• Patterns	• Improve efficiency and provide cost savings

FIGURE 9.7
Understanding our moral framework.

OK, this is a massive exaggeration of the moral framework of a GIS coordinator. It is a little tongue in cheek, but I am *shouting from the rooftops* for all to understand that we go about selling the GIS the wrong way. We must focus on the *outcomes*, rather than the technology. How about never mentioning the word GIS in a sales pitch? Just mention that this technology can do the following:

- Save lives
- Save time
- Promote economic growth
- Engage citizens
- Introduce new ways of thinking

You can have all the passion you want, but if you are not singing from the same song sheet as decision makers, you are doomed to fail.

I do believe that we are defined by the beauty of our language, the rhythm of words, and our ability to communicate with each other. Apart from the ability to daydream about the future, language and communication is the one thing that really separates us from all other species (not including, of course, the remarkable communication skills of the African grey parrot). Using your language with an originality and freshness will set you apart from others. Though conversation is no more than the expulsion of air from the larynx, it can be beautiful and exquisite. The rhythm of words can be seductive and thrilling.

I do not want to come across as a snob, but selling GIS is not like selling used cars or encyclopedias. Not that there is anything wrong with that. GIS has panache, it has sophistication, and it is technical. Language is the only tool that you have to articulate the benefits of GIS.

Language influences actions and creates emotion. There are certain words that can be used effectively to sell an enterprise GIS. Before I give you a helping hand with the language of sales, I want to communicate that there can be loopholes in everything that we have discussed so far in this chapter. We can thoughtfully consider all the factors and components that go into selling, including the obvious and the hidden forces in the effect of sales, but this must be turned into a well-crafted and understandable sales pitch that engages the correct moral framework, hits the heart of the technology, and shows an incredible understanding of the situation.

9.2.5 Selling Topic Five: A Loophole in Our Sales Strategy: The Seven Keys to GIS Success

The *seven keys to GIS success* was developed 25 years ago and is as valid now as it was then. The credit goes to Mr. Curt Hinton, my partner with Geographic Technologies Group (GTG). Not only did he develop a sound sales strategy; he also effectively used the power of seven and bypassed all the noise that is associated with sales. He created an understandable and repeatable methodology that I am guilty of modifying and bastardizing over the years. It is a clear, concise, and relevant message that transcends time and place. Curt introduced the seven keys in the 1980s. Figure 9.8 illustrates the following factors:

1. *Key 1:* Have a well-thought-out enterprise strategic plan—master plan
2. *Key 2:* Have an independent GIS administrator—coordination or governance
3. *Key 3:* Show success quickly and frequently—show success
4. *Key 4:* Explain the uses of the technology frequently—education
5. *Key 5:* Make it useful and easy to use—easy to use
6. *Key 6:* Delegate—do not do all the work—enterprise-wide
7. *Key 7:* Be able to explain and quantify costs versus benefits—return-on-investment (RoI) analysis

As you can see from the seven keys to GIS success, these original points have morphed into something different but similar. The absence of one or more of the keys will greatly reduce the effectiveness of an enterprise-wide GIS. The process of developing a GIS program and the ongoing management of a successful GIS program for an organization must include each of the seven keys: (1) the development of a *master plan*, (2) an effective *coordination or governance* strategy, (3) show *quick success*, (4) a strategy for *education* and training, (5) make it *easy to use*, (6) implement a true *enterprise-wide* solution, and (7) *quantify the benefits and costs.*

9.2.6 Selling Topic Six: Language and Sales

The importance of language cannot be underestimated. The following is a list of terms that I recommend that you use during any GIS meeting. Figure 9.9 shows a list of powerful adjectives that should be used during any sales pitch.

We all have our own success stories to tell, and I hope that the list in Figure 9.9 helps you create your own GIS sales vocabulary. The key to success is to combine these words into phrases that resonate with decision makers. There are three factors that we need to look at. They are entirely separate but need to be tied together through the strategy of sales and marketing.

1. The language of the GIS coordinator
2. The language of sales
3. Moral framework and motivation of elected officials

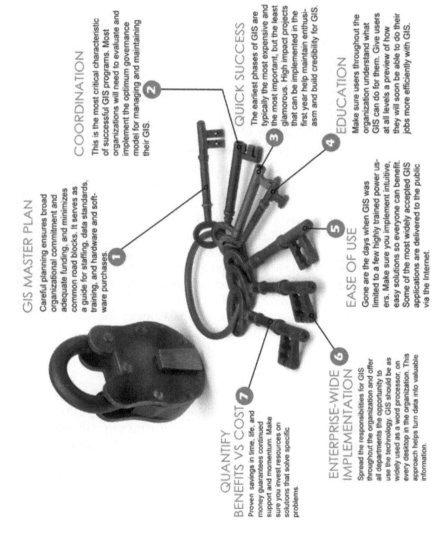

GIS MASTER PLAN

Careful planning ensures broad organizational commitment and adequate funding, and minimizes common road blocks. It serves as a guide for staffing, data standards, training, and hardware and software purchases.

COORDINATION

This is the most critical characteristic of successful GIS programs. Most organizations will need to evaluate and implement the optimum governance model for managing and maintaining their GIS.

QUICK SUCCESS

The earliest phases of GIS are typically the most expensive and the most important, but the least glamorous. High impact projects that can be implemented in the first year help maintain enthusiasm and build credibility for GIS.

EDUCATION

Make sure users throughout the organization understand what GIS can do for them. Give users at all levels a preview of how they will soon be able to do their jobs more efficiently with GIS.

EASE OF USE

Gone are the days when GIS was limited to a few highly trained power users. Make sure you implement intuitive, easy solutions so everyone can benefit. Some of the most widely accepted GIS applications are delivered to the public via the Internet.

QUANTIFY BENEFITS VS COST

Proven savings in time, life, and money guarantees continued support and momentum. Make sure you invest resources on solutions that solve specific problems.

ENTERPRISE-WIDE IMPLEMENTATION

Spread the responsibilities for GIS throughout the organization and offer all departments the opportunity to use the technology. GIS should be as widely used as a word processor, on every desktop in the organization. This approach helps turn data into valuable information.

FIGURE 9.8
The seven keys to GIS success.

REAL SALES LANGUAGE

☑	Faster	☑	Enhance	☑	Valuable
☑	Better	☑	Innovative	☑	Unique
☑	Efficient	☑	Rapid	☑	Extra
☑	Effective	☑	Save	☑	Convenience
☑	Reduction	☑	Identify	☑	Direct
☑	Prioritization	☑	Boost	☑	Original
☑	Minimize	☑	Easy	☑	Smart
☑	Accelerated	☑	Beneficial	☑	Reliable

FIGURE 9.9
Language—key sales words.

Figure 9.10 illustrates how GIS coordinators must change their approach and language to be successful.

9.2.7 Selling Topic Seven: Selling Local Government Data

What we are about to discuss has very little to do with sales and the sales technique but has everything to do with the value of GIS and the philosophical conundrum of selling that data. Currently, there is much discussion about local governments recovering the cost or even generating revenue from creating and disseminating GIS data for public and private consumption. Let me give you the answer to this question before we go through hundreds of hours of discussion or pages of philosophy in search of alternative solutions to this problem.

The solution to this problem, in whatever form it takes, is ultimately a political decision. There is the answer, *it is a political decision that will be answered by the elected officials.* Your job will be simply to educate key staff on the issues and conventional wisdom.

Local government will receive requests from private citizens and businesses for GIS data, products, and services that an organization has produced. The question is, *"What approach should a local government take in responding to this type of request?"*

There are three possible alternatives:

1. Distribute GIS data, products, and services free of charge (zero costs)
2. Recover the cost required to respond to the request (cost recovery strategy)
3. Recover more than the costs listed in number 2, such as the following:
   - Full cost of producing the data or full cost of GIS products and/or services (revenue generation)
   - A percentage of the total cost determined by the organization (revenue generation)

SELLING YOUR ARGUMENT USING THE CORRECT FRAMEWORK

How is it done? A GIS Coordinator Lingo & Mindset	What is sales merit? Positive Sales Language	Why is it done? Elected Official/Decision Makers Moral Framework
I buffered layers to produce a new...	Safer	Equality and community equity
I intersected two layers to produce a new meaning...	Attractive	Strong economy growth
I overlayed data and...	Efficient	Fiscal Responsibility
I calculated the difference between...	Enhance	Improve health of citizens
I produced new and informative statistics that showed...	Effective	Effective welfare of citizen
I analyzed the data and determined...	Transparent	Safer society
We predicted failure at this point...	Innovative	Stronger organization
Our inspection of the data...	Reliable	Cleaner municipality
I used Boolean Logic to determine...	Equality	Effective and well-run organization
We performed geometry analysis to calculate...	Responsible	Provide citizen Access to local government data
I identified the area that would be...	Improve	Allow the organization to provide Quality public service
We performed network analysis and...	Stronger	Empower stakeholders
I created new tools to visualize.	Accessibility	Promote good steward of conversation
Databases analysis and geography produced...	Empower	Engage and notify citizens
We identified patterns of concern...	Promote	Improve efficiency and provide cost savings
I determined the best route using...	Engage	Foster feelings of personal security
I developed the search criteria and identified...	Secure	Secure environment
I performed a spatial overlay analysis and...	Delivers	Reliable mobility
We used real-time response data to...	Maintained	Preserves and promotes historical resources
Our mobile tool set allowed us to determine...	Effective	Promotes cultural enrichment
We performed a statistical analysis on...	Sustainable	Enhance recreation and leisure opportunities
I predicted the incidence of crime and...	Reliable	Sustainable and Reliable infrastructure
We used accurate and reliable data to...	Preserves	Effective use for resources
We modeled the network and...	Foster	Well designed and Maintained attractive livable community
Logistical analysis isolated the problem...	Sustainable	Delivers responsible service
We created informative and graphic rich...	Reliable	Open and Transparent government
I created a three-dimensional model of...	Preserves	Innovative government
I collected field data using the latest...	Fosters	Efficiently run government

FIGURE 9.10
Selling your argument using the correct framework.

When a client asked me for advice about selling GIS data and services, I answered that, in my good judgment, a cost recovery strategy should be used. However, I do want you to know that I recently recommended that a data-sharing agreement include a *good-faith* one-time payment that represented an estimate of between 5% and 10% of the organization's capital outlay on the GIS program. This payment should be seen as the organization recovering the *marginal or a portion of the total costs of the total investment*. This was an exception to the rule.

10

Conclusions

Between the idea and the reality, between the motion and the act, falls the shadow. Between the conception and the creation, between the emotion and the response, falls the shadow.

T.S. Eliot

10.1 Introduction

I am going to break a habit of a lifetime and actually ask for something. I do hope that you do not actually recoil in horror when I make the following requests: please God, do not let me die in the local garden center without attempting to make a difference in someone's life. Please do not let me lose the will to live before I finish this book, please let me be authentic and real, and, for heaven's sake, please do not let me set back relations between the United States and the United Kingdom to the levels that are unseen since the Revolutionary War.

Would it be too much to ask to please let me lighten the shadows of the mundane reality of many GIS professionals? There are many ills in this world that can depress the human spirit. When a GIS coordinator is seen as a doomed hero, or a hopeless character, it can destroy a person's will. Please let me make GIS coordinators into local government gods. Nothing great can be achieved without great suffering. Self-doubt is the inverse of ambition.

I can hear an inexperienced GIS manager saying, "Speak on, dear friend..." What I am attempting to do both in this book and in my life in the geospatial technology industry is something that is fundamentally human. I hope that this book makes a difference in another person's life because, as we all know, the smallest things we do often have the largest impact on others. Every day, we see people act unbelievably kind toward strangers. They do not note the infinity of small kindnesses in the daily news. In short, if in any way this book helps you think about, manage, communicate, or simply feel better about the magnificent world of geospatial sciences, then I will have done my job.

I do not know the hearts and minds of all the GIS professionals in local government. I cannot see the roads of the future. But what I, and now you, have is a vehicle to propel you toward achievement. Admittedly, my formula for success is not a Formula One racing car but perhaps something more like a classic, vintage Austin Martin. I am dreaming here, but GIS can do that to a man. It is so entwined in the day-to-day existence in the twenty-first century that we have already stopped noticing its manifestations, yet its potential remains so limitless.

Geospatial technologies are in your pocket, providing the maps on your smartphone and allowing for delivery orders and car services. They form a key structural component in our skyscraping future with smart technology. It is a thrilling time.

We define ourselves and our world with language. We literally think in words. This is me with the obvious game again, but language is our greatest tool, and thus it would have been spectacularly unfair for this book to tie you up in GIS jargon.

In Section 10.2, I have laid out a brief recap of the book. Remember, refer to the vocabulary list in Appendix B if you ever get confused with the terminology. I hope that the formula for success helps you as much as it has helped me.

10.2 The Importance of Strategic GIS Planning

Developing a plan for your GIS will allow your organization to chart a course for accomplishing all its goals. It is absolutely vital to the organization. The actual goal of the GIS strategic plan should be to dramatically improve the sustainability, endurance, and enterprise nature of the GIS. There are three simple phases to consider:

1. *Phase I: Needs assessment*—This is where you conduct an objective review of the organization's current GIS capabilities and resources.
   A. *Step one*: Online questionnaire
   B. *Step two*: Kick-off meeting and technology seminar
   C. *Step three*: Departmental interviews (Gap analysis and strengths, weaknesses, opportunities, and threats [SWOT] analysis)
   D. *Step four*: GIS needs assessment findings
2. *Phase II: System design*—This is where you make intelligent recommendations for the hardware, software, staff skills, and governance structures that are appropriate for an enterprise GIS.
   E. *Step five*: System design

3. *Phase III: Implementation plan*—This is your opportunity to create a step-by-step implementation plan that details how the system design will meet the goals of the organization.

    F. *Step six*: Business plan and return-on-investment (RoI) analysis

    G. *Step seven*: Implementation plan

There are seven steps, described in Chapter 9, that give you an understandable road map to develop a plan. It does not get any easier than this. Having said that I have also introduced a GIS checkup that may also help you rapidly assess your situation and chart your future. I would also advise you to research, educate yourself, and understand the following:

- The history of GIS adoption in local government
- The challenges, barriers, and pitfalls of trying to implement an enterprise GIS
- The real GIS opportunities in local government
- GIS governance and management issues
- How to turn data into meaningful information, and how we can think differently about the relationship between data
- Infrastructure technology and GIS architecture
- Training, education, and knowledge transfer
- The business and RoI case for GIS
- How you are going to sell GIS within your organization

Figure 10.1 illustrates some important factors of the planning process.

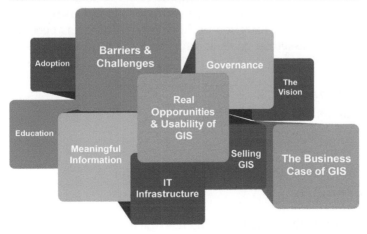

FIGURE 10.1
Understanding strategic planning and successful implementation.

The only way you will become an expert and succeed is to read Chapter 7 and understand the requirements of becoming an effective leader with modern-day skills and creative leadership qualities. Take the harder road, and it will pay off.

10.3 A Possible Formula for Success

Local government officials are an unbelievably diverse and varied group, and it is critically important that they are understood and approached earnestly. *Elected officials* are less interested in the particulars of geospatial technology, but are more concerned with the tangible results of geospatial technology and how it can deliver a return on taxpayer dollars and improve government transparency. *Government administrators* oversee the implementation and management of a GIS initiative. They do everything from managing the GIS budget to organizing seminars and directing operations. Information technology (IT) directors and GIS coordinators meet the challenges of connecting users with GIS technology. There are some great books that describe their different personality styles and their strengths and weaknesses. Understanding this can often be the difference between success and failure.

The reason this book came into being is that I found myself taking apart GIS piece by piece in an attempt to understand it, explain it, and ultimately sell it. What happened after I took it apart is that I had to put it back together again. This is where I stumbled across the formula for success. This formula is the heart of this book as it gives us so much flexibility with understanding the technology and all of its components. The idea behind it consists of compartmentalizing all of the key components and using them to *grade* and benchmark an organization on how successful it has been at meeting or exceeding the essential goals of a successful enterprise GIS. The formula focuses on the six component areas, which are listed as follows:

1. GIS governance
2. GIS digital data and databases
3. Procedures, workflow, and integration
4. GIS software
5. Training, education, and knowledge transfer
6. Infrastructure

This approach allows us to see the heartbeat of an organization and easily determine gaps and opportunities. Figure 10.2 illustrates the outcome of this approach. It is really easy and can be used by any GIS coordinator or IT

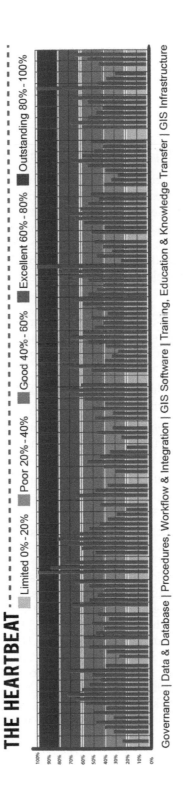

FIGURE 10.2
A GIS heartbeat.

director. Please note that I have introduced a grading system from limited success (0%–20%) to outstanding success (80%–100%).

10.4 What about the Major Obstacles Along the Way?

It is important to shine a bright light on the reasons why local governments often fail to create a truly enterprise, sustainable, and enduring GIS. I wanted to introduce a new vocabulary into the GIS planning process that would allow us to understand and articulate the components, obstacles, and problems of GIS implementation. I am hoping that it helps you communicate clearly and concisely about the process. There are five main GIS strategic planning components, obstacles, or problems that are discussed in this book that can hinder progress:

1. *Strategic GIS components*: Strategic components are the things that are methodically planned, deliberate, and calculated to further the *bigger picture* of the activity of a GIS initiative.

2. *Tactical GIS components*: Tactical components are the things that are considered, prepared, designed, and scheduled to meet the organizations' goals and objectives.

3. *Technical GIS components*: The technical components include the actual mechanical, scientific, procedural, and specialized parts of the GIS implementation process.

4. *Logistical GIS components*: Logistical components comprise everything that is related to the rational supply and support of the technical tools and procedures.

5. *Political GIS components*: Political components are the governmental and ethical components of GIS implementation.

I look closely at the specific challenges, barriers, and pitfalls of attempting to adopt, implement, and manage an enterprise GIS. *The challenges to an enterprise GIS* are often strategic and tactical in nature. They tend to include issues that are associated with the implicit demand for proof, an objection, a query, a test, or a dispute about the validity of GIS. *The barriers to an enterprise GIS* tend to be technical and logistical in nature. *The pitfalls to an enterprise GIS* refer to the hazards along the way. They refer to a hidden or unsuspected difficulty in the implementation process.

In Chapter 4, I have painstakingly documented every possible challenge, barrier, or pitfall that you may encounter along the way. Your job is to understand them, appreciate them, and ultimately *head them off*.

10.5 How Important Is Developing a GIS Vision and Goals and Objectives?

Most organizations lacking a GIS vision usually fail at implementing an enterprise solution.

Without a vision, goals, and objectives, it is impossible to have performance measures. Without performance measures, you cannot monitor progress. Without quantifiable and demonstrated progress, we are doomed.

Developing a vision should be an organic process and culturally specific. It should be creative and analytical. It should involve focused discussions about the experiences, ideas, and perceived function of the organization. A crystal-clear vision statement that details the steps that are necessary for an organization to achieve its vision should be your primary goal as a GIS coordinator. After all, if you succeed in accomplishing this task, everything else you are charged with should fall into place. A clear vision, supported by the necessary steps and the energy, belief, and enthusiasm to complete these tasks, is your single most important task. Do not lose sight of your goal. It is crucial to avoid distraction from the step-by-step actions plans that are outlined in Chapter 5. If you need an outline of how to create a vision, goals, and objectives, here it is:

- *Step one*: Understand the existing GIS situation
- *Step two*: Conduct Blue Sky SWOT analysis
- *Step three*: Build consensus and buy-in from all stakeholders
- *Step four*: Align the GIS with the vision of the organizations
- *Step five*: Create a GIS vision statement, goals, and objectives
- *Step six*: Develop performance measures, outcomes, and metrics

10.6 The Maturing and Evolution of GIS Management

GIS governance models and management styles are often clumsy and awkward. However, they are changing throughout the local government industry. Organizational structures continue to mature and evolve into solutions that can support enterprise and sustainable GIS solutions. In Chapter 6, I focus on explaining GIS governance within local government by discussing three simple model concepts. My experience leads me to believe that local government organizations deploy one of the three main types of governance models:

1. *A centralized model,* where all GIS tasks, except data viewing and analysis, are handled by a central GIS department or division.

2. *A decentralized model,* where GIS data updating and maintenance responsibilities are assigned to individual GIS-participating departments. This used to be called a departmental GIS. There was always an expectation that this governance model would mature into a more enterprise structure. However, no one ever really detailed how this would happen.

3. *A hybrid model,* where GIS is managed and coordinated by a central authority (centralized characteristics), but GIS use and custodianship are decentralized. There is dual accountability and a subject matter expert within this model. This may still be called a corporate or enterprise solution.

Towns, cities, and counties are all at the different stages of GIS growth, and with it, they use different governance models. Evidence suggests that a hybrid model supported by a geographic information officer (GIO) is trending in some of our largest organizations. A hybrid solution with a GIS coordinator, supported by service level agreements and memorandums of understanding, may be trending in mid-size to large organizations. We must also consider the regionalization of GIS, which refers to a situation where organizations in the same geographic region integrate their resources and pool data in order to maximize results. This is also a real topic of conversation throughout local government.

Governance and governance models are a key ingredient for enterprise GIS success. There is a distinct lack of research, understanding, and education.

10.7 A Paradigm Shift: We Need to Think Very Differently about the GIS Coordinator and Enterprise GIS Training, Education, and Knowledge Transfer

Historically, GIS software solutions have driven the training, education, and what little knowledge transfer have existed within local government. As the geospatial industry evolves, and with it, a maturing of the first generation of GIS managers and coordinators, there needs to be a new way that we think about educating, training, and transferring knowledge. It is universally accepted that the GIS coordinator is crucial to a successful GIS implementation process. Training, education, and knowledge transfer solutions are critical for the ongoing success of the GIS. It, therefore, becomes very important to understand what we mean by *GIS training,*

GIS education, and GIS knowledge transfer within local government. (This is me taking apart the components before putting them back together.) Defining exactly what each one does will help us plan around the needs of the future GIS coordinators and, more specifically, the needs of a future workforce.

A paradigm shift is occurring because we are shifting from the software solutions style of education to a future holistic management style that includes the following:

- How to become an effective GIS leader using the principles of deliberate practice
- The modern-day skills of a GIS coordinator using some very human traits
- Creative leadership

What I am saying is that as the GIS becomes such a taken-for-granted technology, we must look to the more human side of GIS management: that of leadership, encouragement, creativity, empathy, and team building. After all, I do believe that local government is an incubator for original and innovative ideas. This future scenario must be supported by a practical and thoughtful review and detail of the characteristics of a future GIS coordinator, as well as the training, education, and knowledge transfer requirements for a future workforce.

10.8 GIS Cost–Benefit Analysis, RoI Analysis, or Value Proposition: Are We Poorer for It?

I said at the start of Chapter 8, *"Your greatest weakness as a GIS coordinator is failing to measure the value of GIS."* I will tell you again that I have literally traveled to every corner of the United States and worked with hundreds of local government organizations. Without exception, no organization has ever attempted to measure or even promote the real tangible and intangible benefits or value of GIS. OK, maybe a couple.

It is invaluable and useful to understand the differences and similarities between the most common buzzwords in the industry, such as cost–benefit analysis, an RoI, and a value proposition. I do not really care which approach you take to quantify the value of GIS, just do it. There are 19 RoI categories in this book that you can use to explain, document, and detail real-world examples of how GIS benefits your government organization. We as a GIS community must start talking about how GIS technology can save lives, inform and notify the public, prevent local governments from being fined, improve the efficiency of virtually all departments within local government, eliminate duplication, and predict events and infrastructure failures, in addition to improving management in all areas of the organization. Your current life is probably the sum of your habits. Let us make evaluating the benefits of GIS a hard habit to break.

10.9 Is There an Art in Selling GIS to Local Government?

There are literally thousands of books and Websites that can show you the basic principles of selling. However, I did not find too many papers or strategies that had a game plan for selling to a group of local government-elected officials or city managers. In fact, most Web sources are geared toward a person who wants to sell anything to anybody. Unfortunately, that type of advice would not help us.

Before we discussed the art of selling GIS to local government in Chapter 9, I stated that "GIS is immensely sound and rich with benefits. It is laden with extraordinary examples of how it supports our world. It has made and continues to make a dramatic and positive difference in the way that local governments operate in the twenty-first century." It would seem that I truly believe in this technology. Remember, people do not buy what you do; they buy what you believe, that is, they buy the idea of thinking differently. People buy from people who enhance their belief that change is possible. If you truly believe in something, clients, employees, friends, and others will too.

I believe that to sell GIS within the local government, we need to understand why small towns, large cities, and mid-size counties invest in the GIS, because if we know why they purchase GIS solutions, we are much closer to developing a sales strategy. It is also imperative that we examine both the *obvious* and *hidden* forces that shape how the *decision makers* in local government think, feel, and behave. I would never underestimate these forces. The obvious forces, including empirical evidence that proves beyond a shadow of a doubt that GIS will benefit the organization, or illustrating how other organizations' best business practices evidence the benefits of GIS technology, or reliably benchmarking your organization against other similar organizations, are key to selling GIS. These hidden forces are not to be taken lightly or, as is too often the case, completely ignored. If we want to talk seriously about selling GIS in local government, these forces require address.

The importance of language cannot be underestimated. We all have our own success stories to tell, and I hope that you can use the information detailed in Chapter 9 to help you create your own GIS sales vocabulary. It is important to combine these words into phrases that resonate with decision makers. There are three factors we need to look at. They are entirely separate but need to be tied together through the strategy of sales and marketing. They are the language of the GIS coordinator, the language of sales, and the moral framework and motivation of elected officials. The idea postulated in this chapter being that selling your argument using the correct framework allows you as a GIS coordinator to change your approach and language.

10.10 The Future of GIS Technology: How Does It Impact the GIS Coordinator or GIO?

Oh, that a man might know
The end of this day's business ere it come!

William Shakespeare, Julius Caesar

Even though we are notoriously poor at remembering the past and equally as poor at forecasting the future, we should, nonetheless, try to anticipate our future self GIS world. So let us turn our attention to the future of GIS, as seen through the eyes of a man who believes that we need to do the following:

- Improve our strategic planning methodology
- Compartmentalize GIS into its natural components
- Understand every possible challenge we may face
- Always develop a vision, goals, objectives, and performance measures
- Create a new strategy for training, education, and transferring knowledge
- Quantify the benefits of GIS
- Learn how to sell GIS technology

Do you remember when on May 2, 2000, President Bill Clinton ordered the Department of Defense to discontinue selective availability—a tool that intentionally degraded Global Positioning System (GPS) signals. Overnight, with the flip of a switch, civilian GPS users were enabled with high-precision accurate GPS. About 100-ft. GPS errors were reduced to 20 ft. This revolutionized the civil and commercial industry. Think about what this did for the world. GPS is pervasive and global, found in everything from cell phones and car navigation systems to watches. By using GPS, you know exactly where everything is, including friends and family. (Please note that my wife and kids asked me to stop tracking [stalking] them on my iPhone).

The reason I am mentioning this milestone in technological advancement is that it has had such a profound influence on our GIS world. The explosion of innovation associated with GPS almost always includes GIS software. Local governments use GPS and GIS to inventory and track infrastructure, track and locate people, and use automated vehicle location technology to manage and monitor their fleet. Radio-frequency identification (RFID) is becoming commonplace in local government. RFID uses electromagnetic fields to automatically identify and track tags that are attached to objects such as garbage containers. So, what can we expect in the future? *Will another switch be thrown?* Most importantly, how will all of these innovations affect the role of a local government GIS coordinator and the team of specialists?

It boggles my mind how geographic location and geographic knowledge have changed the way we think, function, and interact with each other. To understand what will happen in the future of GIS, we should, as I told 100 sixth graders last week, look elsewhere to find the answer. After all, the history of GIS has been a function of the growth of many related disciplines. This pattern continues.

Let me start by saying that the *Internet of things* will impact us all because the speed of connectivity and the sharing of data and information continue at an incredible rate. We have entered a smart connected world whereby your refrigerator, motor vehicle, home alarm system, and even your biorhythms are part of the smart virtual ecosystem. Virtually every device you currently purchase is Wi-Fi enabled and has a way to communicate and report back to headquarters, whether you like it or not.

Our current revolution is about our relationship with technology. It continues to change our life. So, how does this impact local governments and the GIOs? Let us take a look at a list of the future technological advancements that will affect the activities and roles of a GIS coordinator in local government. Figure 10.3 documents possible future innovations that will impact local government operations.

We have now entered this new world of advanced spatial analytics. I believe that you can now lay the foundation for planning, designing, and deploying a sustainable, enterprise, and enduring local government GIS. You are armed with the knowledge and understanding of all the challenges and barriers that will prevent you from absolute success. You are only limited by your management failings, not your economic situation. You must start thinking differently and embracing creative and innovate leadership. Understand what motivates people and the fact that we are influenced by the oddest of things. Approach your goals with a growth mindset. Follow the principles of great performance. You have the tools and techniques to shape your community, and you have the power and genuine belief in GIS to sell this idea. You must align the GIS with your organization's goals and objectives. Create tremendous competitive advantage by anticipating the future of GIS technology.

As a final note, I just want to mention that Captain James Cook was one of the world's most celebrated eighteenth-century mariners whose uncharted voyages of exploration were the most extensive in the world. After completing this book, my goal for July 2016 is to hike to Cook's last stop in English

The Future of Technology
THE INTERNET OF THINGS
INCREDIBLE HYPER PRECISE SPATIAL DATA AND GPS ACCURACY
GIS IN THE CLOUD
IMPROVED DATA SHARING AND WEB SERVICES
ADVANCED CITIZEN ENGAGEMENT
ACCURATE AND INFORMATIVE IN-CAR NAVIGATION
PARTICIPATORY GIS
GEO-SENSORS
INDOOR GPS AND GIS
PASSIVE SOCIAL MEDIA GIS
MINIATURIZED GPS DOTS FOR TRACKING AND MONITORING MOVEMENT
USE OF DRONES FOR REAL-TIME DATA CAPTURE AND MONITORING
IMPROVED DISPLAY TECHNOLOGY
3D AND 4D VISUALIZATION
VOICE RECOGNITION AND GIS
INCREASED HOMELAND SECURITY – GPS DEGRADATION
INNOVATIVE MOBILE GIS
OPEN SOURCE SOLUTIONS AND OPEN GOVERNMENT
ADVANCED CROWDSOURCING
PERSONAL PRIVACY ISSUES
BIG ANALYTICS, GEOSTATISTICS, AND GIS
GPS SECURTIY – SPOOF ATTACK
ADVANCED GEOFENCING AND SOCIA MEDIA
INTEROPERABILITY AND ENTERPRISE INTEGRATION

FIGURE 10.3
The future of technology.

Bay, AK before meeting his untimely death in Hawaii in 1779. The reason I tell you this is that it was said that Captain Cook's constant maxim and practice was never to wait in port for a fair wind. Instead, it was to put to sea and look for one, a characteristic of English seamen. Other sailors never left port till the wind was settled. If you missed my hidden message, do not wait for a fair wind. We live in an era that is rich with opportunity.

References

Alter, Adam L. 2013. *Drunk Tank Pink: And Other Unexpected Forces That Shape How We Think, Feel, and Behave*. New York: The Penguin Press.

Ariely, Dan. 2008. *Predictably Irrational: The Hidden Forces That Shape Our Decisions*. New York: Harper.

Baghai, Mehrdad, Stephen Coley, and David White. 1999. *The Alchemy of Growth: Practical Insights for Building the Enduring Enterprise*. Cambridge MA: Perseus Book Group, p. 32.

Campbell, Heather, and Ian Masser. 1995. *GIS and Organizations*. London: Taylor & Francis.

Colvin, Geoffrey. 2008. *Talent Is Overrated: What Really Separates World-Class Performers from Everybody Else*. New York: Portfolio.

Colvin, Geoffrey. 2015. *Humans Are Underrated: What High Achievers Know That Brilliant Machines Never Will*. New York: Portfolio.

Gilbert, Daniel Todd. 2006. *Stumbling on Happiness*. New York: A.A. Knopf.

Robinson, Ken, and Lou Aronica. 2009. *The Element: How Finding Your Passion Changes Everything*. New York: Viking.

Appendix A

Strategic Planning Clients

Athens–Clarke County, GA—Mary Martin, GIS coordinator

Bahamas Water & Sewerage Corporation, Nassau, Bahamas

Bureau of Land Management (BLM), Anchorage, AK

Calvert County, MD—Kathleen O'Brien, GIS coordinator

Campbell County, WY—Cathy Raney, GIS coordinator

City of Blacksburg, VA—Katherine Smith, GIS coordinator

City of Blue Springs, MO—Gail Porter, GIS analyst

City of Boynton Beach, FL—John McNally, IT director

City of Carlsbad, CA—Karl von Schlieder, GIS coordinator

City of Casper/Natrona County, WY—Michael Szewczyk

City of Champaign, IL—Fred Halner, IT director

City of Chesapeake, VA—Peter Wallace, IT director

City of Cocoa, FL—Keyetta Jackson, GIS coordinator

City of Danville, VA—Inez Pollack, GIS coordinator

City of Dayton, OH—Steve Hill, GIS coordinator

City of Eagan, MN—Tami Maddio, GIS coordinator

City of Edina, MN—Jennifer Bennerotte, GIS coordinator

City of Folsom, CA—Ramona Navarete, GIS analyst

City of Fort Pierce, FL—Marjorie Gaskins

City of Goldsboro, NC—Jeff Cooke, GIS coordinator

City of Goose Creek, SC—Chick Foster, project manager

City of Guelph, Ontario, CN—Chris Sambol, GIS coordinator

City of Hagerstown, MD—Scott Nicewarner, IT manager

City of Hoover, AL—Melinda James Lopez, IT director

City of Johnson City, TN—Lisa Sagona, IT director

City of Kinston, NC—Howard Creech

City of Kissimmee, FL—Tony Curtis, GIS manager

City of Lexington, NC—Brad Benson, IT director

City of Midland, MI—Tony Foisy, GIS coordinator

City of Oviedo, FL—Darlene Jordan, GIS coordinator

City of Pasadena, CA—Jonathan Robinson, GIS coordinator

City of Pearland, TX—Mike Maters, GIS coordinator

City of Rio Rancho, NM—BJ Gottlieb, GIS coordinator

City of Roseville, CA—Scott Adrian, GIS analyst

City of Roswell, GA—Patrick Baber, GIS coordinator

City of South Bend, IN—Shawn Delahanty, IT director

City of Titusville, FL—Rick Story, IT director

City of Unalaska, AK—Erin Reinders, Deputy City manager

City of Valdez, AK—Lisa von Bargen, community development director

City of Virginia Beach, VA—Rob Jessen, GIS coordinator

City of West Hollywood, CA—Francisco Contreras, project manager

City of West Sacramento, CA— Drew Gidlof, GIS coordinator

City of West University Place, TX—Gary McFarland, IT director

City of Wilson, NC—Jimmy Taylor, Engineering

City of Winston–Salem/Forsyth County, NC—Dennis Newman, manager

City of Woodland, CA—Daniel Hewitt, IT/GIS analyst

Columbia County, GA—Mary Howard, GIS manager

County of Brant, Ontario, CN—Bill Leonard, information manager

Duplin County, NC—Tom Reeves, IT manager

Forsyth County, GA—John Kilgore, GIS manager

Gwinnett County, GA—Barry Puckett, GIS coordinator

Halifax County, NC—Doris Hawkins, GIS coordinator

Indian River County, FL—William Rice, GIS manager

Moore County, NC—Lisa Beal, GIS coordinator

Orange County, CA—Quazi Hashmi, project manager

Orange County, NC—Jim Northrup, IT director

Richland County, SC—Pat Bresnahan, GIS coordinator

San Luis Obispo County, CA—Susan Pittaway, project manager

Town of Boone, NC—Chris Miller, GIS coordinator

Town of Branford, CT—Peter Hugret, IT director

Town of Davie, FL—Irene DeGroot, GIS coordinator

Town of Leesburg, VA—John Callahan, GIS manager

Town of Nags Head, NC—Allen Massey, IT director

Town of Windsor, CA—Debbie Shannon, Manager

Wayne County, NC—Chip Crumpler, GIS coordinator

Wicomico County/City of Salisbury, MD—Frank McKenzie, GIS coordinator

Yuma County, AZ—Brian Brady, GIS coordinator

Groerville Utilities Commission (GUC)—Sean Hawley and John Little

Appendix B

Acronyms

AGOL	ArcGIS Online
BBPs	best business practices
BPA	blanket purchase agreement
CAD	computer-aided design
CBA	cost–benefit analysis
CIO	chief information officer
CIP	Capital Improvement Plan
COTS	commercial off-the-shelf
CSD	conceptual system design
ELA	enterprise license agreement
ERP	enterprise resource planning
GASB34	Governmental Accounting Standards Board
GIO	geographic information officer
GIS	geographic information system
GIS SIP	Geographic Information Systems Strategic Implementation Plan
GPS	Global Positioning System
GUC	Greenville Utilities Commission
IT	information technology
KPI	key performance indicator
LGIM	local government information model
MDL	master data list
MOU	memorandum of understanding
ROI	return on investment
SERUG	Southeast Regional User Group
SLA	service level agreement
SME	subject matter expert
SOP	standard operating procedure
SWOT	strengths, weaknesses, opportunities, and threats
VP	vantage point, or vice-president

Index